Risk Management Practices in Construction

Ayodeji E. Oke • Pelumi E. Adetoro
Seyi S. Stephen • Clinton O. Aigbavboa
Lukman O. Oyewobi • Douglas O. Aghimien

Risk Management Practices in Construction

A Global View

Springer

Ayodeji E. Oke
Federal University of Technology Akure
Akure, Nigeria

Seyi S. Stephen
University of Johannesburg
Johannesburg, South Africa

Lukman O. Oyewobi
Federal University of Technology Minna
Minna, Nigeria

Pelumi E. Adetoro
Malawi University of Business
and Applied Sciences
Blantyre, Malawi

Clinton O. Aigbavboa
University of Johannesburg
Johannesburg, South Africa

Douglas O. Aghimien
De Mont Fort University
Leicester, UK

ISBN 978-3-031-35556-1 ISBN 978-3-031-35557-8 (eBook)
https://doi.org/10.1007/978-3-031-35557-8

This Springer imprint is published by the registered company Springer Nature Switzerland AG
The registered company address is: Gewerbestrasse 11, 6330 Cham, Switzerland

To God
who made all things beautiful

Preface

Over the years, the construction industry has been a major driver of physical and economic growth in countries worldwide. Nevertheless, in comparison with other notable industries, the construction industry has always been plagued with risk and uncertainty due to its complexity, scale, and period of completion. In as much as risk management entails forecasting the unforeseeable, it is regarded as the most significant management technique for addressing/dealing with project uncertainty. Risk management can be viewed as a necessary component for adding value to a project by enhancing its cost, time, and quality performance.

Efficiencies and effectiveness of project risk management are sought to be maximized in all approaches to the discipline. Regardless of how the concepts of risk procedures differ based on the particular project, the general risk management concept consists of three basic components: identification, analysis, and response. When managing risk, it is mandated to identify, describe, understand, and assess for treatment. A risk management process that does not result in the adoption of measures to address the risks that have been identified is incomplete and ineffective. The ultimate aim, rather than simply analyzing risk, is to manage it effectively. Risk management in project delivery entails grasping and implementing. It simply provides a methodical approach to thinking about risk and determining how to deal with it.

The book is divided into three parts, the first part is the general introduction that details the background to risk management practices as it pertains to the construction industry. Also, it contains the objective of the book as regards ensuring implementation of effective risk management practice in construction project enhancement and delivery. The second part gives vivid information about risk management practices in some selected developing countries such as Malaysia, Nigeria, Qatar, Saudi-Arabia, South Africa, Sri-Lankan, and Tanzania. Each chapter in this section begins with background introduction and concludes with an outline of some notable points after a definite examination of different written literature. There are references provided at the end of each chapter for further reading or information on the subject matter. This is also the same with the third part of the book, which details countries in developed nations such as Australia, Canada, Sweden, and the United States.

This book is expected to be of good use to the following set of readers: construction professionals, scholars, researchers, government agencies (which include ministries and arms of the government that are involved in the decision-making on construction projects), policy makers, construction stakeholders and educational institutions especially disciplines in the built environment such as architecture, building, land surveying, quantity surveying, estate management, and urban and regional planning among others, stakeholders concerned with the implementation of science, technology, engineering, and mathematics (STEM) practices, consulting firms across various disciplines in finance, value creation, and facility managers in the architecture, engineering, construction, and operation (AECO) industries, as well as undergraduates and graduates in tertiary institutions with interest in risk management-related policies and practices.

The book can serve as a research guide and framework that will encourage construction professionals to deal with risks associated with executing a project. We hope knowledge would be transferred through succinctly expressed risk management policies discussed in the book and enhance construction projects to be executed within the stipulated duration and contracted budget of lean risk and the whole life cycle maximized through identified risk management practices.

Akure, Nigeria — Ayodeji E. Oke
Blantyre, Malawi — Pelumi E. Adetoro
Johannesburg, South Africa — Seyi S. Stephen
Johannesburg, South Africa — Clinton O. Aigbavboa
Minna, Nigeria — Lukman O. Oyewobi
Leicester, UK — Douglas O. Aghimien

Contents

List of Figures

List of Tables

About the Authors

Ayodeji E. Oke is a Senior Lecturer in the Department of Quantity Surveying, Federal University of Technology Akure, Nigeria, and a Senior Research Associate with cidb Centre of Excellence, Faculty of Engineering and Built Environment, University of Johannesburg, South Africa. With more than 350 publications, his research interest is in sustainable infrastructure management (SIM), emphasizing sustainable construction, value management, quantity surveying, and construction in the digital era.

Pelumi E. Adetoro is a Construction Project Manager and currently a postgraduate student of Infrastructure Development and Management at the Malawi University of Business and Applied Sciences, Blantyre, Malawi, under the Africa Sustainable Infrastructure Mobility Scheme. He is a Building Information Management (BIM) enthusiast. His areas of specialization are academic consultancy, BIM education, and general construction. He is a Construction Cost Engineer with practical experience in costing and managing construction projects.

Seyi S. Stephen is a Construction Manager and a graduate of Federal University of Technology, Akure, Nigeria. He is a Social Psychology and Behavioral Sciences enthusiast. He also has a flair for literature, and his areas of specialization are academic consultancy, psychological education, and teaching. He is a Construction Manager and has authored books including *Sustainable Construction in the Era of the Fourth Industrial Revolution*, published by Taylor and Francis, Routledge; *Value Management Implementation in Construction*, published by Emerald Publication; and *Smart Cities: A Panacea for Sustainable Development*, published by Emerald Publication.

Clinton O. Aigbavboa is a Professor in the Department of Construction Management and Quantity Surveying and Director of the DSI/NRF Research Chair in Sustainable Construction Management and Leadership in the Built Environment and the cidb Centre of Excellence and Sustainable Human Settlement and Construction Research Centre, University of Johannesburg, South Africa. He completed his PhD in Engineering Management and has published several research papers in the area of housing, construction, and engineering management and research methodology for construction students. He has extensive knowledge in practice, research, training, and teaching.

Lukman O. Oyewobi has a background in Quantity Surveying and publishes in the general area of construction projects management and strategic management in construction. He is a Senior Lecturer in the Department of Quantity Surveying, Federal University of Technology Minna, Nigeria. He is a cooperate member of the Nigerian Institute of Quantity Surveyors and a Registered Quantity Surveyor with the Quantity Surveyors Registration Board of Nigeria. Dr Oyewobi has extensive tertiary teaching and research experience in the general area of Quantity Surveying.

Douglas O. Aghimien is a Senior Lecturer in Built Environment at the Faculty of Art Design and Humanities, De Montfort University, Leicester, United Kingdom. He is a vibrant researcher with a keen interest in construction digitalization, smart and sustainable construction, and value management. He has a PhD in Engineering Management from the University of Johannesburg and has published over 80 peer-reviewed articles, book, and book chapters.

Part I
General Introduction of the Book

Chapter 1
General Introduction

The construction industry is complex in nature, and understanding all activities within the industry can be challenging. This is because of the unique approaches, methods, and techniques that relate to each project. Lately, there have been some arguments by some parties about referring to it as an industry or as a sector since it consists of many other industries (Ofori, 2015). This complexity, therefore, gives rise to numerous risks, which have effects on the quality, cost, time, and overall performance of projects. Loosemore et al. (2006) corroborated this by highlighting some of the effects of unmanaged risks on construction projects, which include uncertainties in project outcome, financial problem, liabilities, and ineffectiveness in decision-making. The construction sector is a key contributor to any country's economic and social growth. Therefore, there is a need to effectively manage risk to ensure timely completion of construction projects in good quality and within the proposed budget, resulting in swift and profound infrastructural development.

Historical Overview of Risk Management in the Construction Industry

The 1990s—The Emergence of Risk Analysis in the Construction Industry

In the early 1990s, as the identification and evaluation of factors associated with construction project risks became a focal point of industry concentration, various contractors developed varieties of methods for analyzing and assessing these risks. This resulted in failure of construction works which was measured by reviewing the triple constraints (i.e., time, cost, and quality of work done) (Al-Bahar, 1988). Historically, risk-related issues in construction were resolved through a systematic

A. E. Oke et al., *Risk Management Practices in Construction*,
https://doi.org/10.1007/978-3-031-35557-8_1

approach during the precontract stage to avoid time and cost overruns. This method required identifying likely risk factors, assessing them, and proposing solutions to mitigate their effects (Birnie & Yates, 1991). Thus, the factors affecting the cost and duration of the project were classified as controllable and uncontrollable (Akinci & Fischer, 1998).

The Millennium Systematic Approach

There have been several attempts to examine risk factors critically, and as a result, numerous systematic approaches have been developed. Chapman (2001) classified risks into environmental, industrial, client/investor, and project risks. Shen (2001) further classified risk based on the nature of its occurrence which includes financial risk, legal risk, management risk, market risk, policy risk, and political risk.

Chen et al. (2004) studied cost risk management in west rail project of Hong-Kong and classified risk factors affecting construction costs into resource-related factors, management-related factors, and parent-related factors. Dikmen et al. (2007) investigated the use of fuzzy risk assessment to rate the risk of cost overruns in construction projects. The study developed a fuzzy risk assessment methodology using influence diagrams to illustrate the variables that influence project costs. Zeng et al. (2007) categorized risk factors into four major groups; they are human factor, site factor, material factor, and equipment factor. In the early 2000s, major limitations of risk assessment tools and techniques were highlighted such as the analytical hierarchy process (AHP), probability impact (P-I), fuzzy sets theory (FST), Monte Carlo simulation (MCS), and decision support system (DSS) to give a more advanced approach to comprehending and classifying risk in construction as well as other industries.

The Post-2010s—More Sophisticated Technique Development

In recent years, more sophisticated risk management techniques have been developed. In the study by Rezakhani (2012), risk was grouped into three major types, which include external risk, legal risk, and internal risk. The author further divides external risk into unpredictable/uncontrollable and predictable/uncontrollable, while the internal risk was classified as either non-technical/controllable or technical/controllable.

Goh and Abdul-Rahman (2013) classified the numerous risk factors into five different stages: planning stage, design stage, procuring stage, construction stage, and handing over stage. The authors employed checklists, brainstorming, risk register, sensitivity analysis, and Monte Carlo simulation as the risk assessment tool and technique for the classification. The findings of the research identified brainstorming and checklist as the tools that are most prevalent in practice. There have been numerous attempts by researchers to illustrate the relationship between project risks and environmental complexity using a variety approaches (Lazzerini & Mkrtchyan, 2011). Hwang and Ng (2013) in their findings found that deployment of risk management practice in construction projects is generally low, which was attributed to resources and time constraints, a low profit margin, and the fact that it is not cost-effective. The findings of the study revealed there is correlation between positive implementation of risk management and achieving success in executing small-scale construction projects, i.e., in terms of the quality, cost, and time performance.

Objective of the Book

There have been several publications on the definitions and concepts of risk management in the construction sector by different authors (Kartam & Kartam, 2001; Olsson, 2007; Zavadskas et al., 2010), some of which are journals, conference papers, research books, and textbooks. This book aims not just to define risk management and its concept but also to enhance quality project delivery through effective risk management practices.

The scope of this book is limited to risk management in a few selected countries, as it explains explicitly what risks entail in construction and how construction professionals in the countries formulate and mitigate possible risks in their construction practices. The book also gives an insight into the best practices to effectively manage risk in the construction sector in the selected countries. Several impeding factors to the full adoption and implementation of risk management across these countries are identified so as to provide long-lasting solutions to reoccurring impediments in the construction industry.

The book will further improve the awareness and adoption of risk management among professionals and policymakers in the construction industry and also enhance collaboration among project teams. The book will also enhance the development of policies and regulations that will ensure full implementation of risk management for project delivery within stipulated cost, quality, and time.

References

Akinci, B., & Fischer, M. (1998). Factors affecting contractors' risk and cost overburden. *Journal of Management in Engineering, 14*(1), 67–76.

Al-Bahar, J. F. (1988). *Risk management in construction project: A systematic analytical approach for contractors*. Ph.D. thesis, University of California at Berkeley.

Birnie, J., & Yates, A. (1991). Cost prediction using decision/risk analysis methodologies. *Construction Management and Economics, 9*(2), 171–186.

Chapman, R. J. (2001). The controlling influences on effective risk identification and assessment for construction design management. *International Journal of Project Management, 19*(3), 147–160.

Chen, H., Hao, G., Poon, S. W., & Ng, F. F. (2004). *Cost risk management in west rail project of Hong Kong*. AACE International Transactions, INT.09.1-INT.09.5, China.

Dikmen, I., Birgonul, M., & Han, S. (2007). Using fuzzy risk assessment to rate cost overrun risk in international construction projects. *International Journal of Project Management, 25*(5), 494–505.

Goh, C. S., & Abdul-Rahman, H. (2013). The identification and management of major risks in the Malaysian construction industry. *Journal of Construction in Developing Countries, 18*(1), 19–32.

Hwang, B. G., & Ng, W. J. (2013). Project Management knowledge and skills for green construction: Overcoming challenges. *International Journal of Project Management, 31*(2), 272–284.

Kartam, N. A., & Kartam, S. A. (2001). Risk and its management in the Kuwaiti construction industry: A contractors' perspective. *International Journal of Project Management, 19*(6), 325–335.

Lazzerini, B., & Mkrtchyan, L. (2011). Analyzing risk impact factors using extended fuzzy cognitive maps. *Institute of Electrical and Electronics Engineers (IEEE) Systems Journal, 5*(2), 288–297.

Loosemore, M., Raftery, J., Reilly, C., & Higgon, D. (2006). *Risk management in projects* (2nd ed.). Routledge.

Ofori, G. (2015). Nature of the construction industry, its needs and its development: A review of four decades of research. *Journal of Construction in Developing Countries, 20*(2), 115–135.

Olsson, R. (2007). In search of opportunity management: Is the risk management process enough? *International Journal of Project Management, 25*(8), 745–752.

Rezakhani, P. (2012). Fuzzy MCDM model for risk factor selection in construction projects. *Engineering Journal, 16*(5), 79–94.

Shen, L. Y., Wu, G. W., & Ng, C. S. (2001). Risk assessment for construction joint ventures in China. *Journal of Construction Engineering and Management., 27*(1), 76–81.

Zavadskas, E. K., Turskis, Z., & Tamosaitiene, J. (2010). Risk assessment of construction projects. *Journal of Civil Engineering and Management, 16*(1), 33–46.

Zeng, J., An, M., & Smith, N. J. (2007). Application of a fuzzy based decision making methodology to construction project risk assessment. *International Journal of Project Management, 25*(6), 589–600.

Part II
Risk Management in Selected Developing Countries

Chapter 2
Risk Management Practices in Malaysia

Abstract The construction sector has always been plagued with risk and uncertainty when compared to other sectors due to its magnitude as well as its complex and time-consuming nature. In as much as risk management entails predicting the unforeseeable, it can as well be thought of as the most important management tool for dealing with project uncertainty. Risk management is a crucial component for adding value to a project and improving its cost, time, and quality performance. Efficient risk management can bring great result to project performance by improving the productivity. In Malaysia, however, most construction companies do not practice systematic risk management. This condition has resulted in significant numbers of project failure, cost and time overrun, and poor-quality performance. Therefore, this chapter looked into the existing practice of risk management in the Malaysian construction sector and evaluated the various approaches and tools currently being utilized to manage project risk.

Keywords Construction risk · Efficient risk · Project delivery · Project risk · Risk planning · Sustainable construction

Introduction

The construction industry is a major drive in bringing about physical developments in a country by coming up with projects from planning stage to implementation. The implementation and actual completion of these projects will invariably benefit the people and the country at large and in the process enhance national development and subsequently improving the country's economic status.

The construction business is viewed as a high-risk venture to embark on when compared to any other venture because of the various activities involved that are complex in nature. Since each project has its own uniqueness, procedures, and new techniques for carrying them out, a comprehensive approach is developed in relation to the conditions of the projects. The most important factor in project performance is meeting the cost, time, and quality objectives set for the project. However, risk

A. E. Oke et al., *Risk Management Practices in Construction*,
https://doi.org/10.1007/978-3-031-35557-8_2

can occur in a number of ways leading to cost and time overrun, capital loss, loss of life, environmental degradation, and a variety of other failures if not managed properly and as early as possible (Hamimah et al., 2008). As a result, project success or failure is directly proportional to how the risks involved are envisaged right from the onset of construction and throughout other construction phases.

Risk management is essentially making efforts to prevent the impacts of negative occurrences. It can also be viewed as a way of maximizing opportunities to arrive at an outcome by using the information at hand and from past occurrences. Every initiative we embark on involves risk, and how we respond to and cope with it determines the fate of such initiative.

In the study by Raftery (1994), risk and uncertainty refer to situations in which the result of an action or event is most likely to differ from the projected value. As a result, risk can have severe consequences on construction projects. Risk impacts a project's productivity, efficiency, quality, and cost.

Edwards and Bowen (1998) referred to risk management as a useful technique for dealing with the uncertainties that come with construction projects. The researchers cited numerous examples of projects not being completed on time, in good quality, or within budget due to a lack of effective risk management practice in the project management process. Thus, the success of a project in terms of timeliness, sticking to the original budget, and attaining the desired results depends on the risk management skills of each team member involved. Furthermore, Bakker et al. (2010) pointed out that risk management could help increase earnings. As a result, risk management must be methodical and successful for a project to succeed. This makes risk management a topic that deserves careful attention, especially when it comes to project planning and execution. Nevertheless, the primary purpose of risk management is not to eliminate all risks but to manage them to the lowest possible effect. The goal is to provide a framework that will guide major stakeholders in managing risks effectively and efficiently. From the planning to the closing stages of a project, multiple project management tools and procedures should be used. This involves controlling numerous risks connected with the project at each stage. Risk management is therefore considered a critical component of project management. The concept of risk management appears to be novel in the Malaysian construction sector, with only a few companies and industrial practitioners utilizing its tools and techniques.

Risk Management in Malaysia

The Malaysian construction industry has a poor risk management culture (Hamimah et al., 2008). Risk management identification stage in construction is quite distinct from other industries. The risk associated is primarily determined by the characteristics of the construction project (design, methods, materials, location, etc.) and must begin at the very onset of the project planning.

In Malaysia, contractors identify risk using simple, quick, and inexpensive methods such as checklists and brainstorming sessions. Risk analysis requires adequate experience, training, risk management software, and the assistance of a specialist who can advise on the most appropriate response techniques. Risk response for contractors in Malaysia is concentrated on events with a high chance of occurring and a huge impact. Nonetheless, not all companies conduct acceptable risk management operations such as reporting, reviewing, and monitoring ongoing risk management activity.

Adnan et al. (2008) opined that Malaysian construction practitioners should be more proactive in risk management practices. Moreover, Roshana and Akintoye (2005) confirmed that risk management remains a rhetorical subject in the Malaysian construction industry due to a lack of knowledge. As such, most contractors are hesitant to adopt risk management techniques to lower project operational costs. In the study by Hamimah et al. (2008), it was confirmed that risk management is practiced by Malaysian construction companies with a good reputation, a strong financial standing and a wide range of experience in large-scale construction.

As a result, the Malaysian construction sector continues to face significant risks. This creates the need to identify and evaluate current risk management practices in the Malaysian construction sector, including the processes and various tools/techniques. Also, to ascertain the constraints and impeding practices to the implementation of risk management in the country's construction sector.

Project Risk Management

Risk management is a key concept that must be implemented into project management in order to have a successful project. It aims to improve practice performance while adding value to project execution. Risk management is a systematic set of actions and techniques used to monitor risks, and it is essential for getting value and should follow standard procedures and steps (Chapman, 2001). The first is identifying risk factors that have impacts on the risk assessment of the project, which may arise due to the size, nature, and quality of the project desired. Moreover, as is customary in the construction industry, the contractor plays an important role throughout the entire process by actively participating in the project's physical activities. To ensure projects are delivered successfully, the contractors must be available to manage these risks as early as possible.

Fundamentally, project risk management is simply a process of asking and answering a few straightforward questions:

- What risks can be viewed negatively (threats) or positively (opportunities) as impeding or facilitating the achievement of project objectives? (*Risk identification*)
- Which of the risks is by far the most critical? (*Qualitative risk analysis*)

- Given the probabilistic nature of the outcomes, how could these risks affect the overall project outcome in terms of cost and schedule? (*Quantitative risk analysis*)
- What can be done to alleviate the risks? (*Risk response*)
- After taking action, how did the responses influence the outcome of the project, and what is the state of the project now? (*Risk monitoring*)
- Who should be aware of this information? (*Communication*) (Association of Project Management (APM) (2010), Adnan et al., 2008, El-Karim et al., 2017, Maytorena et al., 2005)

Project Risk Management in Malaysian Construction Industry

According to Siang and Ali (2012), the construction industry in Malaysia can be divided into general construction and specialist trade. The general construction includes residential, nonresidential, and civil engineering construction, while the special trade works involve activities such as electrical works, plumbing and services, metal works, carpentry, tiling, flooring works, and glasswork. Despite accounting for less than 5% of the country's gross domestic product (GDP), the industry is a strong growth driver because of its numerous interconnections with other economic sectors (Construction Industry Development Board Malaysia (CIDB), 2006). Without a doubt, the construction industry activities, which include building construction, road construction, electricity or other transmission lines or towers, pipelines, oil refinery, and other specific civil engineering projects, are undeniably essential to the economic development of Malaysia.

The Malaysian construction industry has been impeded by poor performance and substandard project quality. With respect to numerous faults, the Malaysian government initiated the Building Commission (BC) with the primary aim of identifying these faults and proffering solutions to improve the effectiveness of project delivery and reduce delay and cost overrun. The commission found out that the major faults related to construction projects occur due to ineffective risk management practices which resulted in a rise in project completion costs, delays, and poor work quality. The commission also discovered that the role played by quality management systems conforming to ISO standard 9000 had not been sufficient to reduce the inherent shortcoming in the industry (Simu, 2007).

In the general contract, the contractor is fully responsible for carrying out the construction work, and it is the client's responsibility to make any adjustments or modifications to the design. This enables the contractor to focus on the major task at hand and plan toward all unforeseen event and its likely outcomes. The contractors are most times compensated for delay in works due to design modifications. However, because the clients are aware of these rules, any discrepancy with respect to the contractual agreement tendered by the contractor is seen as a means to request unnecessary claims and attract extra cost. Small-scale projects are most likely subjected to design modifications, hence the need to plan for these uncertainties through proper and adequate measures. Parties to this agreement are strongly advised not to deviate from these agreements. In the study by Hellström (2006), contracts that deviate from the general agreements are not to be covered by insurance. As a result, it is essential that the individuals signing the contracts are made aware of the implications of their actions.

Malaysian Standard Risk Management

The risk management technical committee recommended the MS ISO 31000:2010, which is an identical standard to ISO 31000:2009, which was adopted from the ISO standard and provides principles and guidelines to address risk issues, as discussed in Sousa et al. (2012). The standard provides some risk management processes designed to help limit the occurrence of risk on projects (Department of Standards Malaysia, 2010). These processes are as follows.

1. *Communication and Consultation*

Communication and consultation should begin early in the project, and it should be integrated into all other risk management processes. This process deals with risk, its causes, and its consequences. It is also critical to conduct effective external and internal communication and consultation for stakeholders or participants to understand the rationale for decisions and the reasons why specific actions are required.

2. *2. Establishing the Context*

Establishing the context will enable the organization to state its objectives, identifies the external and internal elements to be considered when managing risk, and establishes the scope and risk criteria for the remaining process.

3. *Risk Assessment*

Risk assessment encompasses the entire process of risk identification, analysis, and evaluation.

- *Risk Identification:* Risk identification is used to identify risks, their areas of influence, sources, and consequences. The purpose of this process is to compile a thorough list of risks based on occurrences that could create, improve, avoid, degrade, accelerate, or hinder the achievement of the set objectives. It is regarded as a critical step because the other processes are dependent on the identified risk.
- *Risk Analysis:* Risk analysis is a process that considers the causes and sources of risk, as well as the positive and negative effects and the probability that these events will occur. It is necessary to identify the factors that influence the outcomes and probabilities.

- *Risk Evaluation:* The risk evaluation process is intended to aid decision-making by allowing the organization to categorize the analyzed risks based on their level of significance and then sort out which risks require attention and how fast it is needed.

4. *Risk Treatment*

Risk management entails identifying and implementing one or more risk mitigation methods. Treatments once applied provide or modify controls. Risk treatment is a cyclical process that includes analyzing a risk treatment by determining whether residual risk levels are bearable, developing a new risk treatment, and assessing the performance of that treatment.

5. *Monitoring and Review*

Monitoring and evaluation should be a defined component of the risk management process, including regular checks or surveillance.

The monitoring and evaluation processes should include all parts of the risk management process in order to:

- Make sure that the measures are effective and efficient in both design and implementation.
- Gather additional information to improve risk assessment.
- Examine and learn from occurrences including the success and failures.
- Identify changes in the external and internal contexts, such as changes in risk criteria and the risk itself which may necessitate a revision of risk treatments and priorities.
- Identifying new risks.

Stages of Risk Management Process in Malaysia

Risk management is a systematic process of identifying, analyzing, and responding to project risks. This process increases the possibility and effect of positive events while decreasing the likelihood and effect of negative events, thereby ensuring project success (Yusuwan et al., 2008). Simply put, risk management is a decision-making process that entails having a thorough insight into an already identified risk and ensuring that appropriate steps are taken to mitigate the effects of such risks and their likelihood of occurring to reduce the risks complexities and increase the likelihood of project success.

In the construction industry, risk management process involves three major stages, and they are risk identification, risk analysis and evaluation, and risk response, as illustrated in Fig. 2.1. A significant improvement in the efficiency of construction project management can be achieved by using the risk management process across these stages. Since risk management process is targeted toward eliminating project risks as much as possible, creating an organized structure will

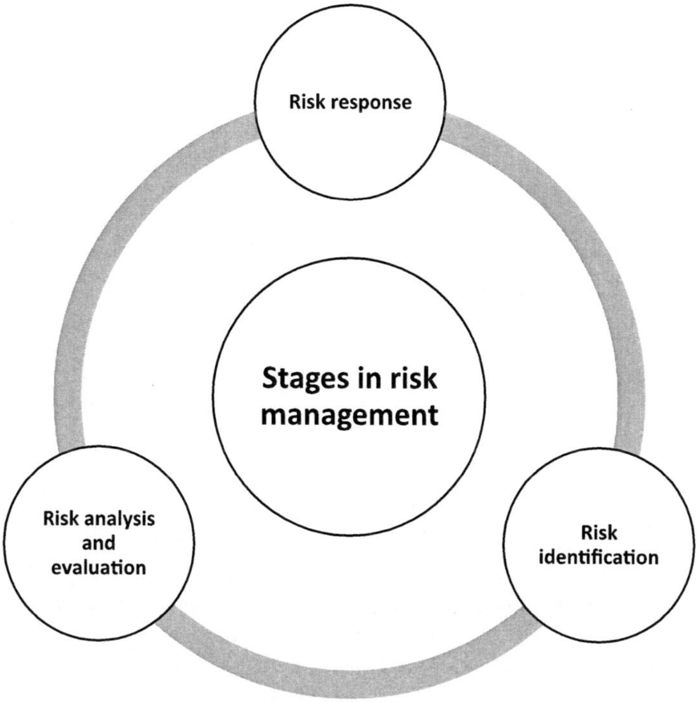

Fig. 2.1 Stages of risk management in the construction industry

allow for quick response to predicted and unforeseen project risks especially when the project is a sensitive one.

Risk Identification

This is the initial phase in risk management that entails detecting, assessing, and classifying the basic impacts of the risks connected to construction projects as well as the interrelationships that exist among these risks in an analytical and continuous manner. The concept of risk identification is widely accepted since it is somehow rare for a project to be discussed without mentioning and identifying possible risks involved in case of failures. It is considered to be a necessary phase since the evaluation and response procedure can only work with the identified potential risks. This is detrimental to the project success as it is mostly done at the onset of project planning.

Risk Evaluation/Analysis

Risk evaluation is described as the process of objectively assessing possible risks, categorizing them by significance, and enabling the top management to choose the most essential ones. The most difficult technique in risk management is risk analysis (Mills, 2007). This is because it entails assessing the likelihood of a risk occurring as well as its impact on the project objectives. Its main goal is to assess risks that have been identified by separating insignificant events, evaluating the likelihood of an unexpected event occurring, and assessing the magnitude of such events. In other words, it is the process of moving from identifying risk to managing it. To assess the possible effects of risk, risk analysis/evaluation incorporates uncertainty in both qualitative and quantitative manners, and thus, risk with high chances or consequences will then be subject to further assessment before decision-making.

Risk Response

Following the identification and analysis of project risks, suitable risk reduction strategies must be applied. These risk reduction measures are primarily focused on the risk existence and potential consequences. A risk process that does not result in the implementation of actions to address the risks that have been identified is incomplete and ineffective. Risk response encompasses project identification, selection, evaluation, and implementation. This phase is designed to improve the degree of risk management and reduce the risk effect and eliminate as much as possible the potential effects. In addition, risk response is centered toward retention, reduction, control, sharing, transfer, and avoidance for an established decision mechanism. The selected response must fit perfectly to address the severity of the analyzed risk. Also, it should be financially viable and feasible in terms of the project timing and be agreed upon by all parties concerned.

Challenges in the Implementation of Risk Management in Malaysia

The Malaysian construction industry faces several risk factors, such as price fluctuation due to inflation, project delay, shortage of skilled workers, and adverse weather conditions among others (Kang et al., 2015). Like its counterparts worldwide, the Malaysian construction industry, also, faces several challenges deterring its proper implementation of risk management practices that will help reduce the occurrence of some of these aforementioned risks. Yusuwan et al. (2008) noted some of these challenges as lack of knowledge, shortage of required expertise, poor cooperation and commitment of project members, lack of proper standards, resistance from management and staff members, among other issues. Similarly, in their assessment of risk management practices in the Malaysian construction industry, Kang et al. (2015) submitted that several challenges are deterring the industry from employing effective risk management. These challenges include a lack of adequate knowledge of risk management, cost and time factors associated with the risk process, reliance on the experience of construction participants, level of education of project participants, and the need for professionals. The challenges deterring the proper implementation of risk management in Malaysia can be summarized into the following:

- Lack of knowledge.
- Shortage of required expertise.
- Poor cooperation and commitment of project members.
- Resistance from management and staff members.
- Cost and time factors associated with the risk management process.
- Reliance on the experience of construction participants.
- Level of education of project participants.
- Lack of proper standards and guidelines.

Benefits of Risk Management in Malaysia

The Department of Standards Malaysia (2010) noted that when the international standard for risk management is implemented correctly, organizations stand to gain several benefits. Some of these benefits are achieving project objectives, encouraging proactive management better, identification of opportunities and threats in projects, better control of projects, and minimizing losses among others. In the same vein, several studies (Ansah et al., 2016; Hillson, 2009; Kang et al., 2015; Loosemore & Cheung, 2015; Siang & Ali, 2012; Ward & Chapman, 2003; Willumsen et al., 2019) have explored the benefits of proper risk management in the Malaysian construction industry. Identified below are some of the benefits of implementing risk management in the Malaysia construction industry. These benefits are identified across various publications.

- It improves project performance in terms of the cost, time and quality.
- It improves the whole process of project management.
- It provides a proactive approach to addressing risks instead of avoidance thereby improving the construction process.
- It improves the organizational culture in terms of organizational learning, resilience, and creating a conducive working environment.
- It reduces financial losses and enhances the continuity of the organization.
- It improves stakeholders and customers/clients' relation.

Conclusion

The construction industry has been developing over the years to meet the demands of the clients. However, the construction industry in countries around the world (Malaysia inclusive) faces several risk factors which if not addressed properly might hamper the successful deliver of projects to cost, time, quality, and the overall satisfaction of the client. To effectively identify and address these risks, the Malaysian industry can do well with further implementation of risk management process into its construction right from the onset of construction and across the post-construction phase. This will help in eliminating challenges that might delay construction, reduce construction quality and so on that at the end of the day will affect project delivery. While there is a risk management standard in place in Malaysia, studies have shown that the proper implementation of these standards is still a problem. This is a result of issues surrounding knowledge and education of construction professionals in relation to risk management among other factors. These challenges identified can therefore be given proper attention to improve risk management within the country's construction industry.

References

Adnan, H., Rahmat, M. N., Marzali, N. F. N., & Jusoff, K. (2008). Risk management assessment for partnering projects in the Malaysian construction industry. *Journal of Politics and Law, 1*(1), 76–81.

Ansah, R. H., Sorooshian, S., Mustafa, S. B. & Duvvuru, G. (2016). Assessment of environmental risks in construction projects: A case of Malaysia. In *Proceedings of the 2016 international conference on industrial engineering and operations management*, September 23–25, Detroit.

Association of Project Management (APM). (2010). *Project risk analysis and management guide* (2nd ed.). Association for Project Management. Retrieved from https://www.perlego.com/book/1595120/project-risk-analysis-and-management-guide-2nd-edition-pdf

Bakker, K. D., Boonstra, A., & Woortman, H. (2010). Does risk management contribute to IT project success? A meta-analysis of empirical evidence. *International Journal of Project Management, 28*(5), 493–503.

Chapman, R. J. (2001). The controlling influences on effective risk identification and assessment for construction design management. *International Journal of Project Management, 19*(3), 147–160.

Construction Industry Development Board Malaysia (CIDB). (2006). *Master plan for occupational safety and health in construction industry.* Retrieved from https://www.scirp.org

Department of Standards Malaysia. (2010). *Risk management – Principles and guidelines (ISO 31000:2009, IDT).* Malaysia Standard. Available at https://medic.usm.my/iso/phocadownload/Malaysian%20Standard%20MS%20ISO%2031000%202010.%20Risk%20management%20-%20Principles%20and%20guidelines%20on%20implementation.pdf. Accessed 29 Jan 2023.

Edwards, P. J., & Bowen, P. A. (1998). Risk and risk management literature in construction: A review and future directions for research. *Engineering, Construction and Architectural Management, 5*(4), 339–349.

El-Karim, M. S., El Nawawy, O. A., & Abdel-Alim, A. M. (2017). Identification and assessment of risk factors affecting construction projects. *Housing and Building National Research Centre (HBRC) Journal, 13*(2), 202–216.

Hamimah, A., Kamaruzaman, J., & Mohd, S. (2008). The Malaysian construction industry risk management in design and build. *Journal of Modern Applied Sciences, 2*(5), 27–33.

Hellström, M. (2006). *Risk manager NCC Group Inc.* Retrieved from https://ww.google.com

Hillson, D. (2009). *Managing risk in projects*. Routledge Publishing.

Kang, B. G., Fazlie, M. A., Goh, B. H., Song, M. K., & Zhang, C. (2015). Current practice of risk management in the Malaysia construction industry – The process and tools/techniques. *International Journal of Structural and Civil Engineering Research, 4*(4), 371–377. https://doi.org/10.18178/ijscer.4.4.371-377

Loosemore, M., & Cheung, E. (2015). Implementing systems thinking to manage risk in public private partnership projects. *International Journal of Project Management, 33*(6), 1325–1334. https://doi.org/10.1016/j.ijproman.2015.02.005

Maytorena, E., Winch, G.M. & Kiely, T. (2005). Construction risk identification. In *Proceedings of the 11th joint CIB international symposium – Combining forces advancing facilities management and construction through innovation*. 13–16 June, Helsinki.

Mills, A. (2007). A systematic approach to risk management for construction. *Structural Survey, 19*(5), 245–252. https://doi.org/10.1108/02630800110412615

Raftery, J. (1994). *Risk analysis in project management*. Routledge Publishing.

Roshana, T., & Akintoye, A. (2005). Process improvement of construction projects in Malaysia: Analysis case studies. In *Proceedings of the 2nd Scottish conference for Postgraduate Researchers of Built and Natural Environment (PRoBE)*. 16–17 November, Glasgow Caledonian University, Scotland.

Siang, L. C., & Ali, A. S. (2012). Implementation of risk management in the Malaysian construction industry. *Journal of Surveying, Construction & Property, 3*(1), 1–15.

Simu, K. (2007). *Risk management on small projects*. Retrieved from https://www.diva-portal.org

Sousa, V., Almeida, N. M., & Dias, L. A. (2012). Risk management framework for the construction industry according to the ISO 31000:2009 standard. *Journal of Risk Analysis and Crisis Response, 2*(4), 261–275.

Ward, S., & Chapman, C. (2003). Transforming project risk management into project uncertainty management. *International Journal of Project Management, 21*(2), 97–105. https://doi.org/10.1016/S0263-7863(01)00080-1

Willumsen, P., Oehmen, J., Stingl, V., & Geraldi, J. (2019). Value creation through project risk management. *International Journal of Project Management, 37*(5), 731–749. https://doi.org/10.1016/j.ijproman.2019.01.007

Yusuwan, N. M., Adnan, H., Omar, A. F., & Jusoff, H. K. (2008). Clients' perspectives of risk management practice in Malaysian construction industry. *Journal of Politics and Law, 1*(3), 121–131.

Chapter 3
Risk Management Practices in Nigeria

Abstract A construction project often necessitates the participation of a large number of individuals with varying skill sets, as well as the coordination of a wide variety of distinct but linked tasks. The particular characteristics of a project and other numerous external uncertainties add to its complexity. With this, construction projects experience cost overruns, delayed schedules, and poor quality due to numerous risks the projects are exposed to at every stage of planning and execution. This chapter examined how risk is managed in Nigeria. In view of that, this study revealed that risk management practice is low and ineffective in most developing nations of the world, and Nigeria is not exempted. This was majorly due to the lack of knowledge on risk management by parties involved in the construction project. This chapter also looked into the best and most effective risk management practices worldwide, which can be used to effectively manage risks in the Nigerian construction industry.

Keywords Construction risk · Project delivery · Project management · Project risk · Risk management · Risk planning

Introduction

Risk management process involves identifying, evaluating, and prioritizing risks, followed by the use of resources in coordinated and cost-effective manner. This is done in order to mitigate the likelihood and resulting effect of unpleasant events, as well as maximizing the realization of opportunities.

In the study by Ojo (2010) on claims and contract disputes in a number of construction projects, it was observed that risks that were not well analyzed or integrated by clients, contractors, or consultants were some of the main causes of claims and disputes in construction projects. It was further stated that risks can occur as a result of several circumstances, including financial market instability, threats from project failures at any stage of the project (i.e., from design to sustainment life cycles), potential lawsuits, credit risk, accidents, naturally occurring catastrophic

A. E. Oke et al., *Risk Management Practices in Construction*,
https://doi.org/10.1007/978-3-031-35557-8_3

events, intentional attack from rival and competitor, and events with uncertain or unpredictable root causes.

In a general view, events in construction can be categorized into two: risks and opportunities (Perry & Thompson, 1992). Positive occurrences are categorized as opportunities, whereas negative events are classified as risks. The Project Management Institute (PMI), the National Institute of Standards and Technology (NIST), Actuarial Societies, and International Organization for Standardization (ISO) have all established a risk management standards, with their methods, definitions and goals varying significantly depending on the application, whether in project management, security, engineering, industrial processes, financial portfolios, actuarial assessments, or public health and safety, among others.

There are several ways of managing threats (risk), these include avoiding the threat right from the onset, reducing the resulting detrimental effect or its likelihood, transferring all or part of the threat to another party, and trying to retain some or all of the potential or actual effects of a specific threat. The opposite of this is applicable when faced with opportunities. To this end, this chapter explores the management of risk in the Nigerian construction industry.

Nigerian Construction Industry

Nigeria has one of the largest economies in sub-Saharan Africa. The economy of the country is supported by the contribution of diverse sectors, including construction. However, the country is challenged by an infrastructure shortage and the deterioration of existing ones. Nigeria's infrastructure stock amounts to just 30% of its gross domestic product (GDP) as of 2021. This is 70% short of the international benchmark set by the World Bank (International Trade Administration, 2021). Considering the astronomical rate of increase in the country's population, it is envisaged that there will be a significant strain on the meager available infrastructure in the nearest future. As a result, substantial investment in infrastructure is needed if the country is to meet its infrastructure needs now and in the nearest future. To deliver this needed infrastructure, the construction industry will play a crucial role.

The construction industry in Nigeria, like its counterpart worldwide, is a significant contributor to the socioeconomic development of the country. The industry contributes about 3.2% to the country's GDP (National Bureau of Statistics, 2020). Moreover, the industry worldwide provides the bases needed by other sectors of the economy to build upon (Giang & Pheng, 2011). Sectors such as transportation, property market, finance, and material production all rely on the product of the industry for their services and contribution to the economy as a whole (Oladinrin et al., 2012). In Nigeria, Olanipekun and Saka (2019) noted that the construction industry makes significant short- and long-term contributions to the economy. It was noted that the industry makes up to 50% contribution to the country's domestic fixed capital formation and up to 20% contribution to employment.

In describing the Nigerian construction industry, Dantata (2008) noted that, like its counter in other countries, the industry in Nigeria could be divided into formal and informal groups. On the one hand, the informal group is unorganized and comprises simple structures delivered by private individuals through the use of private labors. This group makes almost no contribution to revenue generation for the government, and getting statistical data about this informal group is almost impossible. On the other hand, the formal group is organized with registered companies carrying out construction projects through formalized approaches. This group contributes significantly to the growth of the economy, and statistical data are readily available for analysis and possible forecasting. The client of the construction industry includes the government and private individuals/entities. Olanipekun and Saka (2019) have earlier noted that the government is the biggest client of the construction industry, and they are responsible for up to 60% of the industry's output. The private clients cut across oil and gas companies, NGOs, large real estate developers, and homeowners (Dantata, 2008).

Despite its importance to economic growth, provision of employment in the country, and its delivery of projects to both public and private clients, the Nigerian construction industry has been smeared with several issues, such as building collapse, unethical practices, insecurity, poor management, and poor overall performance. Some of these issues emanate as a result of poor risk assessments of projects from the inception to completion.

Risk Management in Nigeria

In Nigeria, effective risk management practices are still in their infancy. Despite being one of the nine knowledge areas set by the Project, Management Institute, the Nigeria construction industry has given little consideration to risk management in the delivery of construction projects (Adeleke et al., 2015). The result of this neglect of risk management is seen in the poor delivery of projects to cost, time, and quality that has characterized the country's construction industry for decades. According to Aibinu and Jagboro (2002), in most cases, participants of projects have little understanding of the risk associated with the project they are handling. As a result, these project participants are unable to put measures in place to ensure these risks are properly addressed and that projects are effectively delivered. Odeyinka et al. (2007) assessed the possibility of occurrences and impacts of certain risk factors in the Nigerian construction industry. The study found that at the precontract stage, design risk, estimating risk, competitive tendering risk, and tender evaluation risk are most likely to occur, and when they do occur, they affect the project adversely. However, at the post-contract stage, financial risk, political risk, contractual risk, logistic risk, legal risk, and environmental risk are more prominent.

Belel and Mahmood (2012) conducted a survey to assess risk management practice in Adamawa State, Nigeria, and identified some of the factors that served as a barrier to risk management practices. They include lack of skilled labor, which appeared to be the major source of risk in construction; and a lack of knowledge, which was recognized as the most intolerant issue. The study further noted that majority of construction professionals are familiar with risk management as it relates to on-site safety hazards rather than those associated with meeting project objectives of cost, time, and quality.

In another study conducted by Aliyu (2013) on risk management in the Nigerian construction industry, level of knowledge of the practice was regarded as the main problem for risk management implementation in Nigeria, with improvement in knowledge being the main driver for development. The study suggested that the best way to improve knowledge is to make risk management a course of study at the higher learning level within the country, as well as to periodically introduce vocational training and risk management workshops.

Project Risk and the Nigerian Code of Cooperate Governance

The Financial Reporting Council of Nigeria, formerly known as Nigerian Accounting Standards Board (NASB), which was signed into law in 2011 with the support of the Ministry of Trade, Industry and Investment, is in charge of providing a code on corporate governance practices and associated listing rules as the country works toward sustainable market (Federal Reporting Council of Nigeria, 2022). This council aims to raise public awareness on important corporate values and ethical practices that will improve the integrity of the business environment and timely implementation of strategic models for the Nigerian market. The codes and conducts are reviewed on a regular basis to reflect the most advanced risks and other challenges facing the Nigerian market.

Managing risks can be more difficult in current times because risk, uncertainty, and threats are not always predictable. The focus of the World Economic Forum tenth edition report (The Global Risks, 2015) was on emphasizing the most significant long-term risks in various categories (economics, societal, geo-political, and technological). The report, based on the conclusions of 1500 audit members, highlighted some of the major risks that pose serious threats to organizations as follows:

1. Uncertainty and volatility.
2. Government regulation.
3. Legal/regulatory adherence.
4. Operational risk.
5. Management and development of talent.

Organizations in Nigeria are not immune to the aforementioned risks. As a result, to the Nigerian Code of Cooperate Governance (2018) provide some risk management principles to help address the risk. These principles are as follows:

Risk Management and Internal Controls The board of directors of the organization is clearly vested with the authority to define and evaluate the risk that is required to meet the organization's objectives, as well as to implement effective risk management and internal controls. They are also responsible for monitoring the introduced systems; therefore, the board is essentially in charge of the design, implementation, and monitoring of the organization's risk management and internal control.

Internal Audit It was recommended in the code that organizations should have an internal audit function and that those that do not have one should review the necessity for this function annually and reveal the results of such review in the cooperate governance report. The provision specifies that the functional role of the internal audit is to conduct an analysis and independent evaluation of the adequacy and effectiveness of risk managed and internal control systems of the organization. Also, the internal audit function may be outsourced and a group with many organizations may share the holding resources to carry out the group's internal audit function. Finally, there is a need to ensure sufficient resources, staff qualifications and experience, training programs, and the internal audit function's budget.

Whistle-Blowing There is a need for the organization to establish a whistle-blowing mechanism to ensure that unethical behavior and violations of established law are reported to the organization's authority for appropriate sanction. It is critical that the whistle-blowing mechanism is reliable, accessible, and secure and that whenever a case is reported, it is reviewed and reported to the audit committee. In addition, if the whistleblower suffers any harm as a result of the disclosure, they are entitled to compensation.

Role of the Audit Committee In reference to the code, risk management is the primary duty of an organization's audit committee, and there is a requirement to monitor its implementation and internal control systems. The audit committee can carry out these obligations, or the organization can create a separate risk committee.

Risk Management Process

Processes in risk management are divided into five steps as follows:

1. Describe the context of event.
2. Identify the risks.
3. Conduct a risk analysis.
4. Treat the risks.
5. Monitor and review treated risks.

Describe the Context of the Event

It is critical to understand the context of the risk occurrence for it to be effectively assessed and managed. The link connecting the source and the environment within which it operates should be defined to clearly understand the parameters for dealing with the identified risks.

Describe the context of event by taking into consideration as follows:

- The strategic context.
- The organization context.

Identify the Risks

The goal at this stage is to figure out the possibility of the event happening and what the consequences (loss or damage) would be if it does happen.

The most important questions to ask here are as follows:

- What is it that could happen? Itemize the risks, events, or accidents that could occur by working your way through each competition, activity, or stage of your event in a systematic manner so as to discover what could happen at each stage.
- What causes it to happen, and how does it happen? Write down all the potential causes and scenarios, as well as a description of the risk, incident, or accident.
- What is the probability of the risk occurring?
- What will be the ramifications if something like this happens?

Risk can either be financial, physical, ethical, or legal. Financial risks are risks that relate to the assets of the organization. It also covers theft, fraud, cost of insurance, membership fees, loans, attendance fees, lease payments, and the payment of compensations or penalties and fines imposed as a result of noncompliance to government laws.

There are also physical risks that relate to risks involving injury to person, harsh environmental and climatic conditions, and the physical assets of the organization, such as landed property, equipment, buildings, vehicles, stock, and facilities.

Ethical risks include real or possible harm to one's image or values, and legal risks include responsibilities imposed on providers, participants, and consumers as a result of laws enacted by governments at the federal, state, and local levels.

Conduct a Risk Analysis

This entails assessing the likelihood and implications of each identified risk, as well as evaluating the risk factors that are most likely to have the biggest impact. These types of risks should be given the highest priority in terms of how they will be managed in the future. To define the priority level of the risk, the risk level is determined by integrating assessments of likelihood and implications.

A risk evaluation method entails the comparison of the level of identified risk through the analytical process and initially established risk criteria and then determining whether risks may be accepted or rejected. These risks may be tolerated with minimal further treatment if they are modest or tolerable, depending on the situation. These risks should be closely monitored and evaluated frequently to ensure that they continue to be acceptable. Once it is determined that the risks do not fit into the low or acceptable category, the situation should be addressed by using some of the treatment options discussed in the next step.

Treat the Risks

Risk treatment entails identifying a variety of treatment alternatives for a given risk, assessing these alternatives, developing risk treatment plans, and putting these plans into action. Treatment alternatives must be carefully considered, with the most acceptable strategy for achieving the desired result being selected from among identified alternatives.

Risk treatment alternatives must be commensurate with the severity of the risk, and the cost incurred in treating it must be proportionate with the possible benefits of treatment.

According to the standard, the following are examples of treatment alternatives:

- *Accepting the Risk*—for example, the majority of individuals would regard minor injuries sustained while engaging in a sporting event to be an inevitable risk of the event in question.
- *Avoiding the Risk*—this involves making a decision about whether or not to proceed with a certain activity or whether or not to choose an alternative event with a tolerable level of risk that matches the objectives of the firm.
- *Reducing the Risk*—it is standard practice to treat a risk by lowering the likelihood of it occurring, reducing its repercussions, or doing both.
- *Transferring the Risk*—contracts or notices are typically used to transfer risk, in whole or in part. Your insurance contract, for example, is probably the most often utilized risk transfer document. The following are some other examples: Lease agreements, tickets, disclaimers, waivers, and warning signs are all examples of transferred documents.
- *Retaining the Risk*—this shows that treatment of risk is not only about risk eradication but rather recognizing the risk and accepting that some risks must be retained. It is critical to examine the amount of risk that is inherent and acceptable in the situation.
- *Financing the Risk*—to finance the risk implies funding the risk effects. This entails providing funds to cover the expenses of putting a risk management treatment plan in place. No matter which approaches you pick to address a risk,

if the risk has been highly assessed, there would be need to take extra care in analyzing the rules, processes, and strategies that will be required to treat the risk effectively. Some of these considerations include the following:

1. What is required to treat the risk?
2. Who is responsible?
3. What is the expected completion date?
4. When a risk has been successfully mitigated or eliminated, how you will know?

Monitor and Review

Just as it is in communication and consultation, monitoring and reviewing are continuous aspects of risk management that are crucial to each phase of the process identified above. This step also encompasses the aspect of risk management that receives the least amount of attention, which results in the risk management plans of many organizations becoming outdated and ineffective over time. By conducting regular monitoring and review, you can guarantee that critical information created by the risk management process is recorded, utilized, and preserved.

There are just few risks that remain constant. As circumstances change, elements that may influence the probability and implications of a result may also change, as well as the factors that influence the acceptability or expense of alternative treatment options, among other things. The review process is an essential component of the risk management treatment strategy.

In addition to being an important part of all key business tasks, risk management should be recognized and managed as such. It is essential that risk management is thoroughly integrated into all operational and managerial activities at all level of the organization and that must be pushed from the executive level.

Risk Management Actions

Risk management actions are measures taken by professionals in order to deal with a variety of risks. Preventive actions and remedial (corrective) actions are the two types of risk management actions that are commonly employed. In construction project, preventive action is used to minimize, prevent, or transfer risks at an earlier stage of the project development, and corrective action is used to mitigate the impact of risks after they have occurred or are about to occur.

Preventive Actions

It has been observed that taking risk management and prevention action at an early phase of the project is more effective. This is emphasized even in early research by Thompson and Perry (1992) where it was concluded that risk management is most beneficial early at the design phase of the project, while there is still the flexibility in design and planning to evaluate how a major risk might be averted. Below are some of the methods that can be used to prevent risk early enough in a project:

1. Make an appropriate time estimates and produce an appropriate program with subjective judgment.
2. Produce an appropriate schedule based on the most recent project information.
3. Make an appropriate time estimates and produce an appropriate program by referring to previous and ongoing similar projects.
4. Plan alternative methods/options as a backup plan.
5. Consciously adjust for bias and add risk premium to time estimation.
6. Make proper time estimation through the quantitative risk analysis techniques.
7. Transfer or share risk to/with other parties.

Corrective Actions

While certain risks that occur as a result of delay in project execution can be mitigated by taking different preventive measures at the early stage of the project, project development is nonetheless delayed during the course of procuring solutions. When there is delay in a project, contractors will be required to take a variety of corrective activities (which is also known as remedial actions) in order to mitigate the consequences of the delay. Some of the corrective actions according to their importance are as follows:

1. Increase the number of labor and/or equipment.
2. Work closely with subcontractors.
3. Ensure strict supervision of subordinates to reduce abortive work.
4. Modify the order of work by overlapping activities.
5. Extend working time.
6. Modify the construction process.

Challenges of Risk Management Practices in Nigeria

In the study by Aliyu (2013), there are some factors responsible for the ineffective risk management practice in Nigeria. Even when the construction professionals in this country are fully aware of this, they have struggled to fully address the situation due to negligence and misconception of the benefits of the practice. In similar vein, Zailani et al. (2019) noted that the issues determining risk management in small construction projects are knowledge and experience related, tight schedules, and construction cost. Lack of knowledge on risk management is stated to be the most prevalent of the abovementioned barriers. Many of the laborers (unskilled) employed for work are not knowledgeable, so there is a need to ensure that they are properly supervised by knowledgeable professionals which are small in number. If supervisors assigned to supervise the unskilled laborers lack adequate risk management knowledge and are unable to identify risk control strategies, they would they be able to transfer practical and theoretical knowledge to the workers.

The challenges to effective risk management in the Nigerian construction industry can be summarized as follows:

- Lack of knowledge on risk management.
- Lack of joint risk management framework by parties.
- Ineffective implementation of risk management.
- Different recognition of risk control strategies.
- Lack of experience of project team.
- Tight project schedule.
- Construction cost.
- Low profit margin of contractors.
- Complexity of risk analytical tools.
- Treating risk management as a one-off activity.
- Underestimation of the impact of risk on projects.

Benefits of Risk Management in Nigeria

The benefits of employing risk management in projects have been explored with a view to encourage construction organizations and participants to adopt proper risk management processes. Okereke et al. (2022) noted that reduction in the occurrence of uncertainties and attainment of project objectives and maximizing opportunities within the business environment are some of major benefits of implementing proper risk management processes. The study by Kpodo (1989) identified some of the inherent benefits of risk management for developing countries, which can be applied to Nigeria as well. The study emphasized how risk management can benefit business organizations, the nation's economy, and individual workers or employees.

- *Business Organization*

  1. Risk management is critical for improving the organization's credit rating.
  2. It is critical for capital utilization.
  3. It is essential for lowering long-term production costs and improve the organization's product price competitiveness.

4. Risk Management helps reduce the effects of natural disaster such as fire, flood etc. that can threaten the survival of the business.

- *Individual Workers or Employees*

1. Risk management ensures worker safety by reducing work-related injuries, which also provides workers or employees with some peace of mind.
2. Risk management ensures that the workforce is productive.

- *Nation's Economy*

1. It improves the competitiveness of the domestic product through effective risk management that automatically improves the country's gross domestic product (GDP).
2. Effective risk management prevents wastage in relation to plant and equipment which saves foreign exchange.
3. Effective risk management will allow the domestic insurance industry to redirect its capacity to areas with higher losses.
4. Ensuring high productivity of workforce will result in an increase in better economic condition.

Conclusion

Risk management implementation is still in the infancy stage in the Nigerian construction industry. This chapter concludes that the limiting factors to effective risk management in the country are knowledge identification from the aspect of the project stakeholders before other construction professionals are involved. With policies embedded in the country's code and regulations, it is a matter of time before risk management practice becomes more adopted not just in the construction industry but also in other industries as well.

References

Adeleke, A. Q., Bahaudin, A. Y., & Kamaruddeen, A. M. (2015). Level of risk management practice in Nigeria construction industry-from a knowledge based approach. *Journal of Management, Marketing and Logistics, 2*(1), 12–23.

Aibinu, A., & Jagboro, G. (2002). The effects of construction delays on project delivery in Nigerian construction industry. *International Journal of Project Management, 20*(8), 593–599.

Aliyu, B. A. (2013). *Risk management in Nigerian construction industry.* Masters dissertation, Department of Civil Engineering. Eastern Mediterranean University, Gazimagusa.

Belel, Z., & Mahmood, H. (2012). Risk management practices in the Nigerian construction industry – A case study of Yola. *Journal of Engineering Sciences, 7*(3), 1–6.

Dantata, S. A. (2008). *General overview of the Nigerian construction industry.* Unpublished masters dissertation, submitted to Massachusetts Institute of Technology.

Federal Reporting Council of Nigeria. (2022). *Financial reporting council of Nigeria announces its intention to early adopt sustainability reporting standards (IFRS S1 & S2) by international sustainability standards board.* Retrieved from https://www.frcnigeria.gov.ng

Giang, D. T., & Pheng, L. S. (2011). Role of construction in economic development: Review of key concepts in the past 40 years. *Habitat International, 35*(1), 118–125. https://doi.org/10.1016/j.habitatint.2010.06.003

Global Risks. (2015). *The world economic forum.* Retrieved from https://www.weforum.org

International Trade Administration. (2021). *Construction sector*. Available on: https://www.trade.gov/country-commercial-guides/nigeria-construction-sector. Accessed 19 Dec 2022.

Kpodo, P. (1989). Risk management in a developing economy. In *Proceedings of the national conference on risk management (FARIM)*, 19–23 March, Nike Lake Hotel.

National Bureau of Statistics. (2020). Nigerian gross domestic product report (Q3). National Bureau of statistics PDF. Available at: www.nigerianstat.gov.ng/. Accessed 24 Nov 2020.

Nigerian Code. (2018). *Nigerian code of corporate governance 2018.* Retrieved from https://www.ecgi.global

Odeyinka, H. A., Oladapo, A. A., & Dada, J. O. (2007). An assessment of risk in construction in the Nigerian construction industry. *Construction in Developing Economies, 107*(1), 359–368.

Ojo, G. (2010). An assessment of the construction site risk-related factors. *The Professional Builder, 1*(1), 10–17.

Okereke, R. A., Zakariyau, M., & Eze, E. C. (2022). The benefits and pitfalls of risk management in construction projects. *CHSM Journal, 2*(1), 58–75.

Oladinrin, T. O., Ogunsemi, D. R., & Aje, I. O. (2012). Role of construction sector in economic growth: Empirical evidence from Nigeria. *FUTY Journal of the Environment, 7*(1), 50–60.

Olanipekun, A. O., & Saka, N. (2019). Response of the Nigerian construction sector to economic shocks. *Construction Economics and Building, 19*(2), 160–180. https://doi.org/10.5130/AJCEB.v19i2.6667

Perry, J. G., & Thompson, P. A. (1992). *Engineering construction risks* (2nd ed.). Thomas Telford.

Thompson, P. A., & Perry, J. G. (1992). *Engineering construction risks: A guide to project risk analysis and risk management.* Thomas Telford, London.

Zailani, M. B., Abubakar, M., & Muhammad, A. (2019). Assessment of barriers to Risk Management (RM) implementation in small construction projects in Nigeria. *African Journal of Built Environment Research, 3*(1), 15–28. https://doi.org/10.33796/ajober.3.1.02

Chapter 4
Risk Management Practices in Qatar

Abstract Risk can be viewed as a crucial issue that must be addressed in all sectors. This chapter provided a broad overview of risks in Qatar and its management, particularly in the construction industry. As construction project planning and execution often necessitates the participation of individuals with varying skill sets coupled with coordination of a wide variety of distinct but linked tasks, the concept of risk assessment is crucial toward achieving the project in sight. The chapter examined some of the tools for effective risk management and how they can be applied in the construction industry. The chapter further identified some of the prevailing challenges as well as the benefits of implementing a risk management process in the Qatari construction industry.

Keywords Construction risk · Project delivery · Risk conceptualization · Risk planning · Risk variety · Sustainable construction

Introduction

Risk assessment in Qatar has been classified into three categories, and they are economic, political, and financial system risks (Humphrey, 2017). The country has experienced a massive surge in infrastructural development over the years as it has become a tourist center for many people around the world. To maintain this revenue stream, the construction stakeholders in the country must be practical in maintaining these infrastructures by considering risks related scenarios right from the project conceptualization to completion.

A. E. Oke et al., *Risk Management Practices in Construction*,
https://doi.org/10.1007/978-3-031-35557-8_4

Qatar Construction Industry

Qatar is regarded as one of the fastest-growing economies in the world (Humphrey, 2017). The construction sector has been a major industry in the development of the economies of most countries around the world, and Qatar is not an exemption. Over the last few years, the Qatari construction industry has undergone a significant change as a result of a shift toward a core number of major infrastructure projects from the ones executed before.

The steady influx of people into the country as a result of the better economy has in turn increased the demand for residential buildings. This has been a dominant reason for the rise in construction activities in the country (Humphrey, 2017). Since the demand is on the increase, construction professionals have responded adequately in terms of putting up major infrastructures in the country.

Furthermore, there is an expectation for a boom in construction activities across the Gulf Cooperation Council (GCC) nations, particularly in Qatar. This is due to the World Cup 2022 that will be held in November 2022. Soccer fans from around the globe will lodge in these facilities over the period of the competition and thus further aid the country in its tourism and recreation approach. Over the next 5 years, it is projected that the construction industry will experience a massive investment of 205 billion US dollars in infrastructure projects such as that to be experienced in roads, rail systems, underground metro, and tunneling projects (Emam et al., 2015).

Risk Management in Qatar

In the management of risk, the following points are essential for consideration:

- Risk origin context (Jaskowski & Sobotka, 2012).
- Allocating process and identity (Hanna et al., 2013; Li et al., 2013).
- Information analysis (Zavadskas et al., 2010).
- Flexible result analysis (Kapliński, 2008; Jaskowski & Sobotka, 2012; Ustinovičius et al., 2010).
- Evaluation and assessment of risk (El-Sayegh, 2007; Ke et al., 2012; Markmann et al., 2013; Skinner et al., 2014).
- Treatment (Ustinovičius et al., 2010).
- Processes of risk (Zavadskas et al., 2010; Kapliński, 2013).
- Monitoring and communication of risk (Zavadskas et al., 2010).

The abovementioned activities tend to reduce threat and increase opportunities during project evaluation. El-Sayegh (2007) and Bon-Gang et al. (2014) examined the diverse risk assessment methodologies employed by different cooperate stakeholder. The result of which discussed the prevalence of risk in every human activity. Since risks associated with construction are numerous and vary in terms of the degree of complexity of the project at hand, whatever chosen risk management practice opted for by the project stakeholders should address necessities concerned with the project.

Several conclusions on risk have associated its effects on any project. These are mostly on the negative side, especially when not discussed and managed as soon as the possible. These outcomes will include monetary loss, damage to property, and injury to person and perhaps a mix of these failures happening at the same time.

When an aspect of risk is identified, it demonstrates that the performance, capacity, potency, quality, and economic cost of a project may all be affected as a result. Planning, design, and development of construction projects are all sensitive to a variety of factors in an unpredictable environment. As a result, construction projects are the product of the actions taken against them in favor of these factors (Jaskowski & Sobotka, 2012). And this automatically affects every aspect of the project outcome irrespective of the corrective measure that might be implemented in the long run.

Risks can be seen in a different form depending on one's point of view; some are categorized based on the likelihood that they will occur, whereas others are categorized on the influence they may have on construction operations. Also, further categorization can be in terms of types and sources. Regardless of how they are classified, they are all intended to achieve a common goal which in turn aids risk management by assisting in formulating risk lists especially when detecting one.

In general, risks can be classified into three major categories. These are known risks, known unknown risks, and unknown risks. Known risks include the smallest number of variations that happens frequently and is inevitable in construction projects. The second category which is the known unknown risks is the one that can be predicted and has a known likelihood of occurrence and impact. The unknown risks include those whose occurrence cannot be predicted with any degree of certainty.

Accordingly, risk can also be classified into groups depending on the sources of concern to the parties involved (Steven, 2017). For instance, there are risks related to time, risks concerned with the environment, risks concerned with cost, and risks concerned with safety. Technical risk, political risk, social risk, economic risk, legal risk, financial risk, health risk, managerial risk, and cultural risk are all subsets of human risks connected with construction projects. And these have even categorized and grouped into 10 classifications, which are as follows:

- *Design*: Defective design, inaccurate quantities, improper design coordination, hasty design, award of design work to unqualified professionals, and discrepancies in contract documents, i.e., drawings, BOQ, and specifications, are all challenges that can develop at the design phase of the project.
- *Physical*: Accidents occur as a result of not adhering to safety measures, the use of substandard materials, the safety of materials and equipment, public safety, and the variation in labor and equipment output.
- *Logistics*: Poor site inspection, unrealistic project schedule, shortage of available labor, resources, and equipment, intense bidding competition, an unclear scope of work, and poor communication between the contractor's home office and project site office accumulate risk during the construction process.

- *Legal*: Non-clarity in work regulations, difficulty in obtaining permits, delay in dispute resolutions, legal issues between the parties to the contract throughout the construction period, and a lack of specialized arbitrators to assist in settling disputes quickly.
- *Environmental*: This includes inclement weather, difficulty in accessing the exact location of the site (located a long distance away), and other environmental conditions (floods, earthquakes, etc.)
- *Construction*: During the construction process, the project is not always implemented in accordance with the specifications due to misinterpretation of drawings and specifications, actual onsite quantities differing from contract quantities, changes to the design, reduction of work quality due to time limitation, hasty bidding, and undocumented change orders.
- *Management*: Issues here include communication gap among the parties involved, confusing strategy as a result of the project complexity, changes in management methods, unavailability of information (which includes uncertainty), and resources allocation.
- *Cultural*: Religion and cultural customs are examples of cultural aspects.
- *Financial*: Delayed payments on contracts, mismanaged cash flow, inflation, the financial failure of the contractor, exchange rate volatility, monopolization of materials as a result of the closure, and other unanticipated political circumstances.
- *Political*: New government acts or laws, inflation, and unpredictable security situations (invasions) are all factors to consider.

Concept/Framework for Risk Management

There are several approaches to managing project risk. However, some of the most often used techniques are described here as follows:

1. Risk identification.
2. Risk analysis.
3. Risk evaluation.
4. Resolving the risks.
5. Risk monitoring and review (Tah & Carr, 2000).

Risk Identification

In identifying risks, brainstorming appears to be the best technique to be used because of its flexible nature. The outcome of risk identification is a complete list of potential risks related to a project. This often comes in the form of a risk register, with management responsibilities for such risks assigned to various risk owners.

Risk Analysis

This particular process involves the selection of risk. It begins with estimating the magnitude of the already identified risk. Here, the causes of each risk occurrence and its possible effects are ascertained. The resulting metrics are then utilized to assess the level of severity of a particular risk occurrence. This is illustrated in the equation discussed by Smith and Merritt (2002).

$$P_{ea}{}^{*}P_{i}^{*}L_{e=}\,L_{e} \tag{4.1}$$

where P_{ea}: probability of risk event
P_i: probability of risk impact
L_e: total loss
L_e: expected loss.

In the Eq. 4.1 above, total loss can also be regarded as the effect in a situation where the risk occurs. The formula for calculating "expected loss" as shown in Eq. 4.1 above is frequently the primary factor in choosing whether to actively manage some risks while postponing action on others.

Risk Evaluation

Risk evaluation is the process of comparing estimated risks to risk criteria in order to establish the risk severity. As previously stated, the key criterion is expected loss since it measures the damage to the project that may be predicted from each risk. The result will be a list of risks that will be actively controlled. The risks and their consequences are all noted in the risk register.

It is essential for a risk map to be developed. This is a graph that shows the correlation between total loss on the horizontal axis (x-axis) and the risk likelihood on the vertical axis (y-axis). The risk map will aid in the prioritizing process.

Resolving the Risks

Risk resolution has to do with the development of action plans for addressing the most severe risk identified in the evaluation process. The main goal of this particular process is to create risk action plans to mitigate the likelihood of a risk occurring and to mitigate its resulting effects if it eventually occurs. Effective risk management does not necessarily engage the risk itself but rather aims to address the risk drivers, i.e., its fundamental truths.

Smith and Merritt (2002) further described the risk resolution process in a more detailed manner. The originators discussed how several beneficial risk resolution and risk reduction practices, such as risk elimination, avoidance, and transfer are detailed in a white paper for energy facility contractors.

Risk Monitoring and Review

This can be seen as a continual effort to verify that the developed risk action plans are moving according to the plan such that unsuccessful plans are withdrawn and any substantial growing or new risks are discovered and addressed quickly.

The primary input is a list of active risks identified for risk management. The findings include modifications to the risk register and listing of new action items for risk resolution.

Concepts of Risk Identification

Risk identification has been seen as a significant process in project management as it helps in identifying the possible risk that might mitigate the successful implementation of a project. Figure 4.1 below shows the summary of project risk identification concepts in regard to identifying risk toward project management. The risk identification process must be all-inclusive as unidentified risks cannot be analyzed, and these unidentified risks emerging at a later stage may have a serious effect on the successful execution of the project (Tah & Carr, 2000).

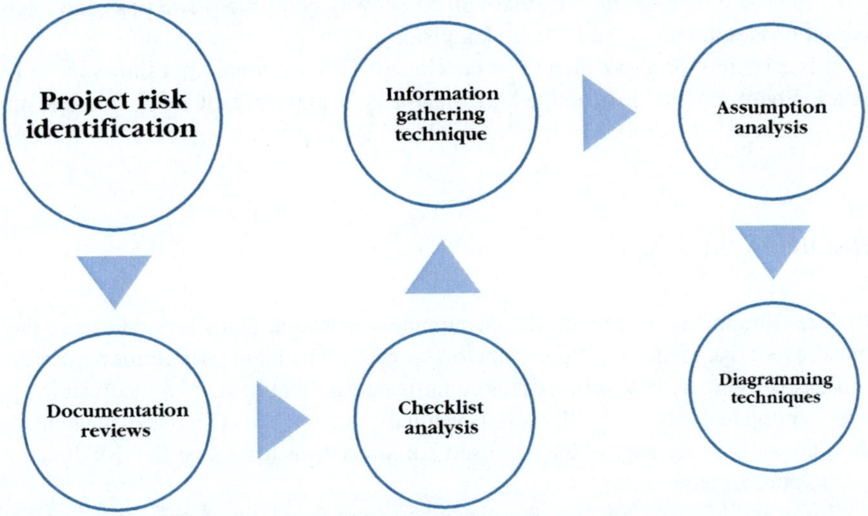

Fig. 4.1 Project risk identification in project

There are several methods and strategies available for identifying project risks. They are as follows:

- Documentation reviews.
- Information gathering techniques.
- Checklist analysis.
- Assumption analysis.
- Diagramming techniques.

Documentation Reviews

This is a process that involves a comprehensive examination of project documents, such as project plans and project files, and the assumptions made. This particular process can identify the sources of risks in a project.

Information Gathering Techniques

Identifying risk necessitates the use of a variety of information gathering strategies. They are as follows.

- *Brainstorming*

This is the most commonly adopted technique in information gathering. Here, the facilitator permits all parties to the project to list risks that might be detrimental to the project without necessarily determining the possibility of their occurrence or effects of such risks. The participants further categorize and analyze these risks before transferring their outputs into a database for more analysis and communication to aid selection.

- *Delphi Technique*

This technique employs quite a number of experts. These experts work on a certain area and are selected from outside the project. They all work independently and transfer whatsoever information they can come up with electronically, majorly through e-mails or printed copies. Each expert is given to offer an anonymous forecast on a certain area. Following that, each expert is anonymously presented with information acquired from others and must make new forecasts based on feedback. The new data are transferred to all experts once more, and the process is repeated until responses begin to converge. This approach is said to be quite successful, although it is time-consuming.

- *Conducting Interview*

Conducting interviews with experienced project participants, stakeholders, and experts is an effort that can lead to risk identification. This approach is regarded as one of the most important sources of gathering data for risk identification.

- *Root Cause Identification*

Root cause identification, as the name implies, involves the process of investigating the major causes of risks in a project. This process allows for the categorization of risks based on their causes. When the fundamental causes of the risk are addressed, effective risk responses may be devised.

- *SWOT Analysis*

SWOT analysis guarantees that the project is examined based on its strengths, weaknesses, opportunities, and threats (SWOT) in order to broaden the scope of assessed risks.

Checklist Analysis

This technique ensures the establishment of a risk identification checklist which is usually based on documented information experienced from a similar project executed in the past. It is difficult to create an all-inclusive checklist; consequently, necessary precautions should be taken to investigate elements that are not included on the previously prepared list. At the completion of the project, it is important that the checklist is reviewed so as to have an improved list that can be used on other projects in the future.

Assumption Analysis

This is a technique for investigating the validity of assumptions in the context of the project. It identifies project risks caused by assumptions that are inaccurate, inconsistent, or incomplete.

Diagramming Techniques

Examples of these may include the following.

- *Cause-and-Effect Diagrams*

They are sometimes referred to as Ishikawa diagrams or fishbone diagrams. These diagrams are useful for determining the causes of risks.

- *System or Process Flowcharts*

These demonstrate how numerous aspects of a system interact, as well as the casual mechanism of the risks.

- *Influence Diagrams*

These are graphical depictions of events that indicate causal influence and event timing. It also depicts the relationships between variables and outcomes.

Dey and Ogunlana (2004) presented an interesting examination of risk management tools and techniques for build operate, and transfer (BOT) projects. The authors emphasized the need for risk identification and risk analysis in such projects.

Challenges of Risk Management in Qatar

One of the most challenging issues in risk management is the inability to apply risk information gained via training or experience to project management tasks. Tah and Carr (2000) provided the following reasons why risk management may not be fully implemented in every project:

- Those engaged are unfamiliar with risk management, and as a result, they are unaware of the benefits it provides.
- Some people dislike being compelled to acknowledge the existence of a risk because it makes them feel vulnerable. Some consider it to be a means of evaluating an individual's performance, while some see it as a call on their competency.
- Risk is costly, and predicting how much money will be generated as a result of it is challenging.
- Companies claim that they lack the necessary time or resources to devote to risk management practices.

Summarily, there are two major factors that have affected the implementation of risk management in the Qatari construction industry over the years; these are cost and time.

- *Time Constraints*: It is understood that the highly pressured environment where construction projects are carried out, coupled with the fast-paced nature of projects resulting in little or no time for risk management practice.
- *Cost Constraints*: It has been frequently discussed that the costs incurred in adopting and administering a proper risk management process are much, thereby making implementation virtually unfeasible. In order to maintain a competitive edge in the industry, organizations must limit the amount of money they spend on personnel to only those who are necessary for the project (who in most cases, do not have the time and technical capacity to carry out risk management).

Benefits of Risk Management in Qatar

In Sousa et al. (2012) in accordance with ISO 31000, an international risk management standard utilized in Qatar, implementing risk management in any organization enhances efficiency, accountability, and the level of trust stakeholders have in the organization while reducing losses and augmenting risk analysis as well as risk assessment capabilities. Some of the other identified benefits of effective risk management for organizations are stated as follows:

- Gain a competitive advantage.
- Reduces costs in a long term.
- Responds to change by offering effective solutions.
- Value creation and protection.
- Increases the chances of the organizational objectives being met.
- Proactively identifies opportunities and threats.
- Identifies and reduces risk across the organization.
- Establishes a standardized basis for making choices and planning.

Conclusion

This chapter covered risk management practices in the Qatari construction industry. Starting from the introduction of the country's construction and the lapses in fully adopting risk management practices due to cost and time, the chapter went further in evaluating risk management practices, barriers, benefits, and other related sections concerned with the country's construction practice. Finally, it can be inferred from the chapter once the construction stakeholders in Qatar can fully comprehend fully the benefits that come with risk management practice, especially with the whole life cycle of massive infrastructures embarked upon by the country, and it will take their construction to even greater summit than what is being experienced now.

References

Bon-Gang, H., Xianbo, Z., & Ping, T. L. (2014). Risk Management in small construction projects in Singapore: Status, barriers and impact. *International Journal of Project Management, 32*(1), 116–124. https://doi.org/10.1016/j.ijproman.2013.01.007

Dey, P. K., & Ogunlana, S. O. (2004). Selection and application of risk management tools and technique for build-operate-transfer-projects. *Industrial Management and Data System, 104*(4), 334–336.

El-Sayegh, S. M. (2007). Risk assessment and allocation in UAE construction industry. *International Journal of Project Management, 26*(4), 431–438. https://doi.org/10.1016/j.ijproman.2007.07.004

Emam, H., Farrell, P., & Abdelaal, M. (2015). Causes of delay on large infrastructure projects in Qatar. In *31st annual ARCOM conference,* September 15–17, Lincoln.

Hanna, A. S., Thomas, G., & Swanson, J. R. (2013). Construction risk identification and allocation: Cooperative approach. *Journal of Construction Engineering and Management, 139*(9), 1098–1107. https://doi.org/10.1061/(ASCE)CO.1943-7862.0000703

Humphrey, A. (2017). *Risk management project.* Retrieved from https://www.academia.edu

Jaskowski, P., & Sobotka, A. (2012). Using soft precedence relations for reduction of the construction project duration. *Technological and Economic Development of Economy, 18*(2), 262–279. https://doi.org/10.3846/20294913.2012.666217

Kapliński, O. (2008). Usefulness and credibility of scoring methods in construction industry. *Journal of Civil Engineering and Management., 14*(1), 21–28. https://doi.org/10.3846/1392-3730.2008.14.21-28

Kapliński, O. (2013). The utility theory in maintenance and repair strategy. *Procedia Engineering, 54*(2), 604–614. https://doi.org/10.1016/j.proeng.2013.03.055

Ke, Y., Wang, S., & Chan, A. P. C. (2012). Risk management practice in China's public-private partnership projects. *Journal of Civil Engineering and Management, 18*(5), 675–684. https://www.researchgate.net/deref/http%3A%2F%2Fdx.doi.org%2F10.3846%2F13923730.2012.723380

Li, H. X., Al-Hussein, M., Lei, Z., & Ajweh, Z. (2013). Risk identification and assessment of modular construction utilizing fuzzy analytic hierarchy process (AHP) and simulation. *Canadian Journal of Civil Engineering., 40*(12), 1184–1195. https://doi.org/10.1139/cjce-2013-0013

Markmann, C., Darkow, I. L., & Gracht, H. V. (2013). A Delphi-based risk analysis – Identifying and assessing future challenges for supply chain security in a multi-stakeholder environment. *Technological Forecasting and Social Change, 80*(9), 1815–1833. https://doi.org/10.1016/j.techfore.2012.10.019

Skinner, D. J. C., Rocks, S. A., Pollard, S. J. T., & Drew, G. H. (2014). Identifying uncertainty in environmental risk assessments: The development of a novel typology and its implications for risk characterization. *Human and Ecological Risk Assessment: An International Journal, 20*(3), 607–640. https://doi.org/10.1080/10807039.2013.779899

Smith, P. G., & Merritt, G. M. (2002). *Proactive risk management-controlling uncertainty in product development*. Productivity Press.

Sousa, V., Almeida, N. M., & Dias, L. A. (2012). Risk management framework for the construction industry according to the ISO 31000:2009 standard. *Journal of Risk Analysis and Crisis Response, 2*(4), 261–275.

Steven, H. (2017). *Qatar construction market outlook 2017–2023* (Technical report: white paper series). Retrieved from https://doi.org/10.13140/RG.2.2.13091.25122

Tah, J. H. M., & Carr, V. (2000). Information modeling for construction project risk management system. *Engineering, Construction and Architectural Management., 7*(2), 107–119.

Ustinovičius, L., Ševčenko, G., Barvidas, A., Ashikhmin, I. V., & Kochin, D. (2010). Feasibility of verbal analysis application to solving the problems of investment in construction. *Automation in Construction, 19*(3), 375–384. https://doi.org/10.1016/j.autcon.2009.12.004

Zavadskas, E. K., Turskis, Z., & Tamošaitienė, J. (2010). Risk assessment of construction projects. *Journal of Civil Engineering and Management, 16*(1), 33–46. https://doi.org/10.3846/jcem.2010.03

Chapter 5
Risk Management Practice in Saudi Arabia

Abstract The Saudi Arabian construction sector generates so much revenue that it is considered the largest public construction sector in the Persian Gulf. The country has experienced massive growth in the construction sector as a result of the government ongoing investment in social and economic infrastructures such as housing, transportation, hotel, and energy infrastructure. However, the delivery of these infrastructures is not free from risks and the attributed risk impacts. Therefore, this chapter offers insight into the causes of risk in the construction process and the management of these risks. The chapter submitted that the major risk factors in the country are cost- and time-related. Furthermore, the chapter suggested that construction clients are to blame for the majority of this problem owing to frequent changes in the scope of work and orders. The chapter concluded by offering new risk assessment tools for advanced construction that have recently become popular in some developed nations of the world.

Keywords Construction risk · Risk analysis · Risk assessment · Risk management · Risk tools · Urban risk

Introduction

Saudi Arabia is one of the fastest-growing economies in the Middle East, with a gross domestic product (GDP) growth rate of 5% (Algahtany et al., 2016). The development of infrastructure in both urban and rural areas of the country has accelerated enormously. According to Algahtany et al. (2016) and Deloitte (2013), Saudi Arabia has the largest public construction sector in the Gulf with $575 billion spent on public building projects between 2008 and 2013. Wanies-Guirgis (2012) noted that the construction industry is expected to reach $610 billion between 2015 and 2020, and this is in addition to the enormous revenue being generated by the country's large oil reserves, which has been a significant boost to the country's economy in recent years.

A. E. Oke et al., *Risk Management Practices in Construction*,
https://doi.org/10.1007/978-3-031-35557-8_5

The government on their part is enacting new laws and processes to further strengthen the economy of the country, one of which is a sustained diversification of exports and investment initiatives. In recent years, the rapid and significant development of the housing sector has been a key contributor to the country's economy growth, and the result has paid dividends greatly to the government and the people at large (Bajwa & Syed, 2020; Dhonte et al., 2010).

Saudi Arabian Construction Industry

The Saudi Arabian construction sector is now increasing at a pace of 4.5% per year, with a total value of $30 billion (Bajwa & Syed, 2020; Deloitte, 2013). The construction sector could expect continuous development in the future years, since the government plans to continue to invest in social and economic infrastructures. The most prevalent forms of construction projects in the country today are transportation, hospitality, residential accommodation, and energy infrastructures. In delivering these projects, the construction industry attracts a significant number of workers, with a 26.4% contribution to employment in the country (The General Authority for Statistics, 2020).

Saudi Arabia's construction projects are distinct in their qualities and the numerous risks they entail. Risks are frequently increasing due to the extended duration of projects, fiscal magnitudes, and the sophisticated organizational structures (Smith et al., 2013). To this end, researchers have put in efforts in evaluating the impacts of risks on the cost of project, health and safety, and time constraints. Also, further evaluations have been done on risk and its management, particularly on construction projects at the design and construction stages (Ebrat & Ghodsi, 2014; Zou et al., 2007). It has been noticed that the existence of several contractual parties (contractors, subcontractors, building designers, suppliers, and non-construction professionals), as well as wide range of cultural differences appear to pose numerous basic risks in the Saudi Arabian construction sector (Taylan et al., 2014).

Risk is an indeterminate and uncertain event which have impacts on project components either positively or adversely (suitability, quality, quantity, time, cost, and safety). Dikmen et al. (2007) stated that it is nearly impossible to eliminate all risks from construction projects; as a result, construction projects require general risk assessment and process management in order to cope with all forms of threats in all situations. This will help in the project whole life cycle and value creation.

Overview of Risk

Risk can be described as the exposure of a certain event to loss or gain or the probability of loss or gain occurring multiplied by the amount of the loss or gain. Events are said to be certain if the likelihood of their occurrence is 100%, and they are said to be completely uncertain if the probability of their occurrence is 0%. Risk has been defined further as something that cannot be predicted accurately or that arises from unexpected sources that result in a negative deviation from the project time and cost estimates (Ward & Chapman, 2003).

Project delay and time overrun are both indicators and consequences of the occurrence of a risk. Every component of a business venture has the potential to be fraught with risk, and this is true in every industry. Some risks are caused by our own actions, while others are inevitable due to situations beyond our control. As a result, stakeholders must always examine the risks that are most likely to be encountered in the course of their daily operations. The first step in the risk assessment process is to identify and categorize all risky situations. Risks linked with the construction industry are any problems that can cause injury or loss to persons, property, or other interests (Jannadi, 2007). It will be necessary to assess and quantify risks when they have been identified.

However, in recent years, varieties of risk assessment systems have been developed. Peltier (2004) demonstrated how a probability/impact matrix might be very effective in evaluating and prioritizing risk levels before taking any action to minimize them. It is especially useful since risks with a higher possibility of happening demand the most attention, yet uncertainty varies greatly. In summary, risk can be defined as an event that has an impact on specific objectives. This impact can be positive or negative and can occur at the micro-, meson, and macro-levels.

Reasons for Risks Issues in Saudi Arabia Construction Industry

The result of several researches carried out in Saudi Arabia over the past years has revealed the critical issues for non-performance, inefficiencies, time, and cost overruns in the construction industry. Time overrun and cost of implementation are regarded as some of the most prevalent and serious issues in the country's construction projects (Assaf & Al-Hejji, 2006, Faridi & El-Sayegh, 2006, Kashiwagi et al., 2015). Assaf and Al-Hejji (2006) revealed that between 60% and 70% of public construction projects in Saudi Arabia had delays in their completion time. According to the study results, the average delay percentage ranged from 10% to 30% of the initial contract period. An assessment of 49 case studies in the country's western province revealed that the average delay in these situations was 39% of the projected duration (Kashiwagi et al., 2015).

Further study by Abdul-Ghafour (2011) estimated that the total value of public projects that have gone past their anticipated completion periods is roughly $147 billion. In addition, Allahaim and Liu (2015) noted that the issue of projects overrunning their initial cost in the Saudi Arabia construction industry has been a longstanding concern for decades. Al-Turkey (2011) discovered that 80% of Saudi construction projects were prone to cost overruns. Despite these problems over cost and time overruns, studies have noted that there are practically no effective risk management strategies in the country construction sector (Baghdadi & Kishk, 2015).

Ikediashi et al. (2014) conducted a study to identify and analyze the factors that led to the failure of infrastructural projects in Saudi Arabia. According to the findings of the study, the most significant cause of project failure was poor risk management practices. Also, Albogamy and Dawood (2015) revealed that there is a significant absence of risk management practices in Saudi Arabia which identify and analyze the effects of risk variables due to the lack of engagement of clients and their agents in the construction process. Traditionally, clients do not accept responsibility for risks and instead automatically shift them to other parties (Al-Sobiei et al., 2005). In the Saudi construction industry practices, the majority of risks are transferred to vendors, with no risk being allocated to the clients (Al-Salman, 2004). Albogamy and Dawood (2015) further discovered that the majority of significant risks faced during the early phases of a project are caused by the clients themselves. Trigunarsyah and Al-Solaiman (2015) added that these risks are linked with insufficient decision-making on the part of clients who lack the necessary knowledge and expertise.

Ibn-Homaid et al. (2011) assessed the sources of change orders in the Saudi construction sector and argued that clients are the primary source of risks based on project scope changes and change orders. This has resulted in an average 11.3% increase in project costs. A similar submission was made by Kashiwagi et al. (2015) who stated that the majority of delays in the Saudi Arabia construction industry were caused by the owners of the projects (clients). Bajwa and Syed (2020) also submitted that delayed payment to contractors and unrealistic project deadlines are the two major risk factors common in the country's construction industry. This gives credence to the earlier submissions of clients being a major source of risk on project delivery in Saudi Arabia.

Risk Management

Risk management refers to the methods used to manage possible risks by identifying, analyzing, and mitigating them to reduce the possibility or effect of unfavorable negative occurrences to maximize the realization of emerging possibilities. The outcome may help reduce the likelihood of a risk arising and the negative impact if one does occur (Tallaki & Bracci, 2005). In recent years, risk management has become a significant component of the project management process. Indeed, Akintoye and MacLeod (1997) have earlier mentioned that the state of the construction industry has necessitated the implementation of risk management and analysis processes.

Project risk management is an important component of project management theory and technique (Project Management Institute, 2008). This is because unanticipated events frequently occur throughout a project (Banaitis & Banaitiene, 2012). Given the significance of project risk management in the functioning of project management, it is predicted that the effectiveness of risk management will substantially impact project performance. An early study by Strutt (1993) to assess the impacts of risk management on project performance discovered that risk management strategies are effective in achieving successful project outcomes.

Risk and uncertainty have been major factors that have caused setbacks to the growth of the construction industry due to the complex, dynamic, and time-demanding nature of the industry. Therefore, risk management is one of the most essential management strategies for dealing with project risks and uncertainties since it is focused on anticipating the unexpected. Risk management strategies are essential for adding value to a project and improving the performance of projects in relation to the cost, time, and overall quality. Despite this, the majority of construction enterprises in Saudi Arabia do not apply risk management techniques, resulting in the failure of construction projects due to cost overrun, schedule slippages, and poor-quality performance. As such, risk management can have strong impact on the project performance especially when enhancement is required along the line of project execution (Jin et al., 2007).

Risk Management in Saudi Arabia Construction Industry

Risks are inherent in any business or projects. In a situation where there is an occurrence of risk, risk management system is implemented with the aim of identifying and quantifying those risks so that a conscious decision may be made on how to manage such risks (Markmann et al., 2013). The Project Management Body of Knowledge (PMBOK) (2004) highlighted risk management as one of the nine focuses in project management and defines it as a systematic method of identifying, analyzing, and responding to risks in a project. This strategy entails maximizing the likelihood and implications of positive events while minimizing the possibility and consequences of events that are detrimental to the project's objectives. This can be done by carrying out risk analysis on the project as early as possible. Some of the various methods (but not limited) of analyzing risks are as follows:

1. Fuzzy TOPSIS bid/no bid model (Ravanshadnia & Rajaie, 2013).
2. Applying TOPSIS-F method in fuzzy environment (Tamosaitiene et al., 2013).
3. Based on intelligent agent (Smeureanu et al., 2012).
4. RAMCAP (Risk Analysis and Management for Critical Asset Protection) by introducing new risk-related factors (Yazdani et al., 2011).
5. Fuzzy Synthetic Model (Abdul-Rahman et al., 2013).

In further discussion on risk management and how it should be implemented correctly, the following factors must be considered.

1. Source of the risk.
2. Identification and allocation of processes.
3. Analysis of information.
4. Analysis of the flexibility of results.

5. Assessment and evaluation of risks.
6. Risk treatment.
7. Risk function or process.
8. Risk monitoring and communication.

A well-designed risk management system stresses the use of a more organized approach, more effective centralized control, and better transfer of risk information between parties. In addition to this, the system increases awareness of the effects of risk on projects. The ability to translate uncertainty into risk and risk into opportunity is essential for effective risk management. As a result, by maximizing opportunity, limiting risk, and lowering uncertainty, the project and organization stand to gain more.

Sundararajan (2004) stated that failure to manage risk events properly would result to severe consequences, such as increase in the cost of the project, modification of the organization's capital structure, delay in the commercial operation of the building or facility, budget overrun, decrease in cash flow, liquidated damages claim from contractor, poor-quality delivery, and rework after the final completion of the project. Other researchers, such as Lemos et al. (2013) and Roth & Espersen (2004), described effective risk management as the successful performance of the whole procedure of risk management, which includes risk management planning, risk identification, risk analysis, risk response, and risk monitoring and control.

Despite all these highlighted benefits and effectiveness that accompany the implementation of a risk management system, in the practical sense of it, it has been discovered that organizations around the world that have successfully implemented risk management are relatively low, with the most optimistic reports placing the figure at less than 25% percent of all organizations that have implemented these processes (Kutsch & Hall, 2010). The Saudi Arabia construction industry is no exception. This poor use of risk management systems is significantly harming the industry considering that construction projects have higher inherent risks due to the participation of various contracting parties, including clients, designers, contractors, subcontractors, and suppliers (PMI, 2004). Clients, contractors, and the public are constantly affected by the consequences of risks not being assessed (Zavadskas et al., 2012). Therefore, it can be said that the construction industry involves a high level of risk which has a significant impact on the parties involved; also, the systematic approach of analyzing and managing risks related to construction has been a major challenge to practitioners in the industry (Kaplinski, 2009).

There are some important factors that are needed to be considered for effective risk management. These factors include the significance of the risks, current risk management strategies used in organizations, the present condition of risk management systems in organizations, and the impediments to successful risk management as seen from the perspective of major stakeholders. Choudhry and Iqbal (2013) researched analysis revealed that the most significant risks are financial and economic considerations, followed by quality as the industry seeks to prevent or transfer these risks.

Risk perception is an essential part of risk management. Furthermore, the attitudes and barriers to risk management and advantages of perceived risks are required for risk analysis and management process.

Risk Assessment

Risk can be assessed in many ways one of which is the use of heat map. Heat map is considered to be one of the important risk assessment tools that can be utilized in construction projects. A heat map is a graphic that may be used to illustrate risks based on certain classifications and priorities. It makes use of standard coded colors to indicate the severity of the risks. For example, the red color represents massive risk, the yellow color represents a moderate risk, and the green color represents extremely low risk (Javalgi et al., 2011). Following the identification and quantification of risks, the appropriate risk response techniques will be identified as the next procedures. The terms abolition, decrement, retention, and relocation are all used to refer to risk management strategies that are routinely employed. The risk response techniques that have been selected are implemented and monitored.

Tools for Risk Assessment in Saudi Construction Industry

Integrated Hybrid Methodology This methodology developed by Taylan et al. (2014) was used to identify and analyze critical risk variables related to construction projects at King Abdulaziz University in Riyadh, Saudi Arabia (KAU). The key risk variables addressed in this case were time, cost, quality, safety, and environmental sustainability. Based on the data provided, this technique employs a relative relevance index to prioritize project risks. This hybrid technique may be used to classify construction projects utilizing the fuzzy AHP and fuzzy TOPSIS techniques. Taylan et al. (2014) attempted to incorporate crucial qualitative qualities into construction project performance assessments, and they translated qualitative data into quantitative metrics. The findings indicated that assessing the overall risks associated with construction projects is possible. Using this method, it is possible to identify the project with the lowest risk, which is quite helpful.

The Fuzzy Risk Management Method Fuzzy risk management method was proposed by Liu (2012). It has the potential to reduce the occurrence of work-related risks. This method uses quality function deployment (QFD) tables to demonstrate the exact relationship between construction substances, classifications, and reasons for risk. The fuzzy analytic network process (ANP) method can be employed for the identification process.

Assessing project risks using the fuzzy implication technique was developed based on the failure modes and effects analysis (FMEA). In the case of work-related hazards, the results of the performance evaluations are obtained and then offered a satisfactory assessment for risk values. A fuzzy multiple criteria decision-making (FMCDM) tool is used to estimate the level of risk associated with several urban construction projects. Wang and Chang (2012) investigated the comparative effects of 20 known risk concerns on project coordination using consistent fuzzy preference relation (CFPR) in this scenario. The fuzzy multiple attribute direct rating (FMADR) determined the chance of each risk factor occurring.

Factors Affecting Risk Management Practices in Saudi Arabia

In research done for Middle East countries which include the kingdom of Saudi Arabia and Jordan, we have different risk factors, which are listed below.

Material-Related Factor This factor shows material-related issues that often cause delays in construction. Examples of material-related issues that can cause delay are material supply and shortage of construction materials, both of which have significant implications as regards the time overrun of the project. Another example is the rise in the price of raw materials which may result in an increase in overall construction costs. The third material quality problem is of serious concern to construction parties. Lewry and Crewdson (1994) stated that construction players must not make any concessions on this issue. Failure to take these material-related factors seriously may have an adverse effect on the overall construction process.

Project-Related Factor This method draws attention to the difficulties associated with the numerous elements and stages of a construction project. Among the five project-related attributes, the first is *"inadequate early planning of the project,"* which is one of the key causes of delay that construction parties do not take cognizance of during assessment (Bramble & Callahan, 2010). The second attribute is *"shortage of equipment availability,"* which refers to a scenario that occurs during the construction process due to improper equipment planning during the project conceptualization phase. When it comes to the third attribute, *"lack of systematic engineering method to identify project time"* occurs as a result of making use of unprofessional and inappropriate construction methods, as well as a possible lack of dedication on the part of the engineer or project manager. The fourth project-related attribute, *"poor performance of the lowest-bidder contractor in the tendering procedure,"* is a relatively prevalent issue that contributes to project delays in Middle Eastern nations. Most public clients in developing nations tend to consider contractors with the lowest bid without considering the bidder's qualifications, experience, and expert knowledge on project management procedures and planning techniques (Agumba & Fester, 2011). Lastly, the fifth attribute, *"low manpower output"* results from poor site supervision or the engagement of an unprofessional/unskilled workforce, which negatively impacts the quality delivery of the construction project.

Contractor-Related Factor This factor reflects delays caused by contractors or subcontractors who lacked the necessary skills, knowledge, experience, and competencies for the completion of the project. Critical contractor-related delay characteristics are poor site management and supervision by the contractor and poor communication by the contractor.

Potts and Ankrah (2008) stressed the necessity of efficient supervision and communication in attaining cost and time performance in construction projects. Another problematic delay factor is that contractors' technical staff may be unable to deal with issues or not grasp the complexities of a project, both of which can result in time overruns. The third problematic delay factor is "poor qualification, skills and experience of the contractor's technical staff." The contractor may lack the requisite qualifications, skills, and experience to execute the work, which may result in time overrun. There is also "shortage of qualified contractors," which is another major cause of delays in the Middle East region. The government often engages foreign contractors to execute the majority of the capital projects in the country since there are not enough qualified contractors. Lastly, "lack of proper schedule and planning of the project by the contractor" might result in inaccurate estimations of a variety of features and event of things that may develop throughout the construction process (Agumba & Fester, 2011).

Owner-Related Factor This particular factor has six underlying characteristics; it illustrates the significance of the client's decisions and actions before and throughout the construction phase. The first characteristic is a "lack of cooperation with contractors involved in the project," which results in construction projects failing to meet their specifications (Doloi et al., 2011). The second feature, "delay in the approval of contractor submittals," explains the situation where a client fails to fulfill his contractual obligations or attends to a request from the contractor. This is simply an indication of a lack of collaboration with contractors on the client's part. The third factor, "changes in the scope of the project," which has been a regular experience with the project managers in the Middle East, is often due to a lack of understanding of the project scope or/and design. This is a common occurrence in the region (Kasimu, 2012). This often results in time delays due to rework, errors, duplication of activities, a lack of drive, and reaching cost limits. The fourth characteristic, "breach or modification of contract by the client," typically refers to the engagement of unskilled or amateur contractors who lack the necessary skillsets (Doloi et al., 2011). The fifth characteristic is "poor qualifications and supervision of owner's engineer," which might pose serious challenges for the owner and the other construction stakeholders to attain a timely completion of the project. Finally, the significance of the sixth characteristics, the "slow decision-making process of the owner," may be evaluated by the rankings of several academics, who have classified it as one of the top five causes driving construction delays in developing nations.

Consultant-Related Factor It indicates that there are problems with the consultant. It has four significant consultant-related characteristics. The first characteristic, "poor qualification and supervision of the staff of the consultant engineer," will have a detrimental effect on the client's objectives in terms of both cost and time. The second attribute, "delay in the approval of working drawings," is frequently caused by a breakdown in communication between the consultant and the approving authority. The third attribute, "absence of consultant's site workers," is important to keep track of this on a continuous basis to minimize the time overruns (Assaf & Al-Hejji, 2006). The last attribute, "inadequate qualifications of consultant assigned to project," highlights the need to hire skilled consultants to implement good project management practices and planning procedures.

Design-Related Factor This emphasizes the significance of design-related errors and changes made by the consultant and client as a result of either a lack of communication, expertise, knowledge, education, and professional experience with the consultant.

External-Related Factor Work delays caused by changes in government laws or unforeseen circumstances such as adverse weather or changes in the external environment are examples of this type of factor. This factor can be further expressed in terms of the following:

- *Poor Communication Between Construction Parties*: Poor communication among parties in the construction has brought about risks. There should be proper discussion and brainstorming on what are the circumstances (challenges) to expect during the construction process, which must be mitigated, so that the client's objectives are achieved.
- *Delay in Decision-Making*: The delay in decision-making from either the client or the designer is one of the reasons affecting management practices in Saudi Arabia. The inability to come to a consensus between the client and the management is causing delays in what to do and what is better for the project.
- *Changes in Management Ways*: Constant changes experienced in management are a contributing factor to why there is little adherence to risk management practices in Saudi Arabia. While one management team might be mindful of implementing risk management practices, another might not.
- *Unreasonable Project Cost Frame*: In executing risk management practice in construction, it costs a lot to put all the sections together. Also, when the budget sum is outrageous or unreasonable, such could limit the zeal to implement risk management practices.
- *Improper Construction Method*: When the wrong approach is used in executing a project, faults and errors will occur at one or many stages of the project. To avoid cost in setting up a risk assessment, some contractors may adopt easier or more convenient construction methods that might not serve project's purpose, size, and nature.

- *Changes in Client Ways*: This is related to constant demand and frequent change initiated by the client. The change might come in terms of design, interior, and method, concerned with an already agreed schedule. This, in turn, affects the risk management process that might have been implemented and accepted based on the initial information.
- *Design Changes and Inappropriate Design*: Risk can only be managed when there is a standard design, but inappropriate changes in design affect risk management practices in Saudi Arabia.
- *Weather Condition*: Weather condition is also part of the factors affecting risk management.

Barriers to Risk Management

Liu (2012) discovered that construction companies lacked the expertise and knowledge required for practical implementation of recognizing hazards, assessing risk, minimizing risks, and preparing for emergencies (RAMP) because it has only been in use for a short period in the industry. The researchers further discovered ineffective coordination between the parties involved, a lack of availability of specialist risk management consultants, and a deadline-driven schedule for construction projects. According to the conclusions of that study, the following seven obstacles to RAMP implementation were identified as follows:

- Inadequate awareness of risk management processes.
- Lack of experience.
- Lack of necessary information about the project.
- Lack of cooperation between parties to the project.
- Availability of qualified consultants with strong knowledge of risk management.
- The costs of implementation.
- Time constraints.

Conclusion

The practice of risk management in Saudi Arabia has been addressed in this study. The tools used to analyze risk management and the reasons for risk issues in Saudi Arabia are among the topics covered. In addition, the study discussed the obstacles and factors that hindered the implementation of risk management in the Saudi Arabian construction industry, which centered on time, knowledge, and cost. The construction stakeholders in the country can do more to educate construction professionals about risk management practices and how these practices can enhance the delivery of construction projects.

References

Abdul-Ghafour, P. (2011). *Projects worth SR550bn stalled, contractors ask government to step in.* Retrieved from https://www.arabnews.com/node/401823

Abdul-Rahman, H., Wang, C., & Lee, Y. L. (2013). Design and pilot run of fuzzy synthetic model (FSM) for risk evaluation in civil engineering. *Journal of Civil Engineering and Management, 19*(2), 217–238. https://doi.org/10.3846/13923730.2012.743926

Agumba, J. B., & Fester, F. C. (2011). Identifying tools and techniques for managing construction project delivery in small and medium enterprises in UK construction industry. *Journal of Social and Development Sciences, 2*(4), 204–213.

Akintoye, A. S., & MacLeod, M. J. (1997). Risk analysis and management in construction. *International Journal of Project Management, 15*(1), 31–38.

Albogamy, A., & Dawood, N. (2015). Development of a client-based risk management methodology for the early design stage of construction processes: Applied to the KSA. *Engineering, Construction and Architectural Management, 22*(5), 493–515.

Algahtany, M., Alhammadi, Y., & Kashiwagi, D. (2016). Introducing a new risk management model to the Saudi Arabian construction industry. *Procedia Engineering, 145*(2), 940–947. https://doi.org/10.1016/j.proeng.2016.04.122

Al-Sobiei, O. S., Arditi, D., & Polat, G. (2005). Managing owner's risk of contractor default. *Journal of Construction Engineering and Management, 131*(9), 973–978.

Al-Salman, A. A. (2004). *Assessment of risk management perceptions and practices of construction contractors in Saudi Arabia: Master dissertation.* King Fahd University of Petroleum and Minerals. Retrieved from: https://www.proquest.com/openview/186f4e1dc43c3f313e43c1827b562c4c/1?pq-origsite=gscholar&cbl=18750&diss=y

Al-Turkey. (2011). *The reality of projects in terms of organization, and structure and the reason for the failure and success in Saudi Arabia.* Retrieved from https://www.alwatan.com.sa

Allahaim, F. S., & Liu, L. (2015). Causes of cost overruns on infrastructure projects in Saudi Arabia. *International Journal of Collaborative Enterprise, 5*(1/2), 32–45.

Assaf, S. A., & Al-Hejji, S. (2006). Causes of delay in large construction projects. *International Journal of Project Management, 24*(4), 349–357.

Baghdadi, A., & Kishk, M. (2015). Saudi Arabian aviation construction projects: Identification of risks and their consequences. *Procedia Engineering, 123*(1), 32–40.

Bajwa, I. A., & Syed, A. M. (2020). Identification of major construction sector risks in Saudi Arabia. *World Transactions on Engineering and Technology Education, 18*(2), 247–2560.

Banaitis, A., & Banaitiene, N. (2012). *Risk management in construction projects.* Retrieved from https://www.researchgate.net/publication/309727885

Bramble, B. B., & Callahan, M. T. (2010). *Construction delay claims* (4th ed.). Aspen Publishers.

Choudhry, R., & Iqbal, K. (2013). Identification of risk management system in construction industry in Pakistan. *Journal of Management in Engineering, 29*(1), 42–49. https://www.researchgate.net/deref/http%3A%2F%2Fdx.doi.org%2F10.1061%2F%28ASCE%29ME.1943-5479.0000122

Deloitte, R. (2013). *GCC Powers of Construction 2013: Construction sector overview.* Retrieved from https://www.deloitte.com/assets/DcomMiddleEast/Local%20Assets/Documents/Industries/Real%20Estate/Construction/me_realestate_gcc_construction_ppt_13.pdf

Dhonte, P., Bhattacharya, R., & Youssef, T. (2010). *Demographic transition in the middle east-implications for growth, employment, and housing.* Retrieved from https://google.com/demographic-transition-in-the-middle-east

Dikmen, I., Birgonul, M. T., & Han, S. (2007). Using fuzzy risk assessment to rate cost overrun risk in international construction projects. *International Journal of Project Management, 25*(5), 494–505.

Doloi, H., Sawhney, A., Iyer, K. C., & Rentala, S. (2011). Analysing factors affecting delays in Indian construction projects. *International Journal of Project Management, 30*, 479–489.

Ebrat, M., & Ghodsi, R. (2014). Risk assessment of construction projects using network based adaptive fuzzy system: An empirical study. *KSCE Journal of Civil Engineering, 18*(5), 1213–1227.

Faridi, A. S., & El-Sayegh, S. M. (2006). Significant factors causing delay in the UAE construction industry. *Construction Management and Economics, 24*(11), 1167–1176.

Ibn-Homaid, N. T., Eldosouky, A. I., & Al-Ghamdi, M. A. (2011). Change orders in Saudi linear construction projects. *Emirates Journal for Engineering Research, 16*(1), 33–42.

Ikediashi, D. I., Ogunlana, S. O., & Alotaibi, A. (2014). Analysis of project failure factors for infrastructure projects in Saudi Arabia: A multivariate approach. *Journal of Construction in Developing Countries, 19*(1), 35–52.

Jannadi, A. O. (2007). Risks associated with trenching works in Saudi Arabia. *Building and Environment, 43*(5), 776–781.

Javalgi, R. R., Deligonul, S., Dixit, A., & Cavusgil, S. T. (2011). International market reentry: A review and research framework. *International Business Review, 20*(4), 377–393.

Jin, X., Doloi, H., & Gao, S. (2007). Relationship-based determinants of building project performance in China. *Construction Management and Economics, 25*(3), 297–304.

Kaplinski, O. (2009). Information technology in the development of the Polish construction industry. *Technological and Economic Development of Economy, 15*(3), 437–452. https://doi.org/10.3846/1392-8619.2009.15.437-452

Kashiwagi, D., Sullivan, K., Greenwood, D., Kovell, J., & Egbu, C. (2015). Source of construction industry instability and performance problems. In *Proceedings of the ASCE construction research congress*, April, San Diego.

Kasimu, M. A. (2012). Significant factors that cause cost overruns in building construction project in Nigeria. *Interdisciplinary Journal of Contemporary Research in Business, 3*(11), 775–780.

Kutsch, E., & Hall, M. (2010). Deliberate ignorance in project risk management. *International Journal of Project Management, 28*(3), 245–255.

Lewry, A. J., & Crewdson, L. F. E. (1994). Approaches to testing the durability of materials used in the construction and maintenance of buildings. *Construction and Building Materials, 8*(4), 211–222.

Lemos, F. F., Johanson, J., & Vahlne, J. (2013). Risk management in the internationalization process of the firm: A note on the Uppsala model. *Journal of World Business, 46*(2), 143–153.

Liu, H. T. (2012). A fuzzy risk assessment approach for occupational hazards in the construction industry. *Safety Science, 50*(4), 1067–1078.

Markmann, C., Darkow, I. L., & Von der Gracht, H. (2013). A Delphi-based risk analysis – Identifying and assessing future challenges for supply chain security in a multi-stakeholder environment. *Technological Forecasting and Social Change, 80*(9), 1815–1833. https://doi.org/10.1016/j.techfore.2012.10.019

Peltier, T. R. (2004). Risk analysis and risk management. *Information System Security, 13*(4), 44–56.

Potts, K., & Ankrah, N. (2008). *Construction cost management: Learning from case studies.* Retrieved from https://www.api.taylorfrancis.com

Project Management Institute (PMI). (2004). *A guide to the Project Management Body of Knowledge (PMBOK guide).* Project Management Institute Incorporation.

Project Management Institute (PMI). (2008). *A guide to the Project Management Body of Knowledge (PMBOK guide).* Project Management Institute Incorporation.

Ravanshadnia, M., & Rajaie, H. (2013). Semi-ideal bidding via a fuzzy TOPSIS project evaluation framework in risky environments. *Journal of Civil Engineering and Management, 19*(1), 106–115.

Roth, J., & Espersen, D. (2004). Risk reflections: based on his extensive experience, David McNamee discusses the components that make up an effective system of risk management. *Internal Auditor, 61*(5), 75–93.

Smeureanu, I., Ruxanda, G., Diosteanu, A., Delcea, C., & Cotfas, L. A. (2012). Intelligent agents and risk-based model for supply chain management. *Technological and Economic Development, 18*(3), 452–469. https://doi.org/10.3846/1392-8619.2008.14.428-443

Smith, N. J., Merna, T., & Jobling, P. (2013). *Managing risk in construction projects*. Wiley.

Strutt, J. (1993). *Risk assessment and management: The engineering approach centre for industrial safety and reliability*. Retrieved from https://www.scholar.google.com

Sundararajan, V. (2004). Risk measurement, risk management and disclosure in Islamic finance. In *Proceedings from the seminar on comparative supervision of Islamic and conventional finance*, 7–8 December, Beirut, Lebanon.

Tallaki, M., & Bracci, E. (2005). Risk allocation, transfer andmanagement in public–privatepartnership and private financeinitiatives: a systematicliterature review. *International Journal of Public Sector Management, 1*(2), 1–23. https://doi.org/10.1108/IJPSM-06-2020-0161

Tamosaitiene, J., Zavadskas, E. K., & Turskis, Z. (2013). Multi-criteria risk assessment of a construction project. *Procedia Computer Science, 17*(1), 129–133. https://doi.org/10.1016/j.procs.2013.05.018

Taylan, O., Bafail, A. O., Abdulaal, R., & Kabli, M. R. (2014). Construction projects selection and risk assessment by fuzzy AHP and fuzzy TOPSIS methodologies. *Applied Soft Computing, 17*(1), 105–116.

The General Authority for Statistics. (2020). *Labor market statistics Q2 2020*. Available from: https://www.stats.gov.sa/en/34

Trigunarsyah, B., & Al-Solaiman, S. (2015). Individual factors influence on client involvement: The case construction projects in Saudi Arabia. In *Proceedings of the 6th international conference on construction engineering and project management*, 11–14, October, Busan.

Wanies-Guirgis, C. (2012). Trade integration in the Middle East and the North Africa region: What is in their horizon following the Arab Spring? *Journal of Current International Trade, 21*(2), 43–53.

Wang, T., & Chang, T. (2012). Application of consistent fuzzy preference relations in predicting the success of knowledge management implementation. *European Journal of Operational Research, 182*(3), 1313–1329.

Ward, S., & Chapman, C. (2003). Transforming project risk management into project uncertainty management. *International Journal of Project Management, 21*(1), 97–105. https://doi.org/10.1016/S0263-7863(01)00080-1

Yazdani, M., Alidoosti, A., & Zavadskas, E. K. (2011). *Economic Research-Ekonomska Istraživanja, 24*(4), 27–40.

Zavadskas, E. K., Turskis, Z., & Tamosaitiene, J. (2012). Contractor selection for construction works by applying SAW-G and TOPSIS grey techniques. *Journal of Business Economic and Management, 11*(1), 34–55. https://doi.org/10.3846/jbem.2010.03

Zou, P. X., Zhang, G., & Wang, J. (2007). Understanding the key risks in construction projects in China. *International Journal of Project Management, 25*(6), 601–614.

Chapter 6
Risk Management Practices in South Africa

Abstract Risk management often deals with analyzing market and non-market risks (short- and long-term risks). Also, risk management analyzes the influence of risks on the corporate environment by developing strategies to mitigate risk exposure. As a developing economy, South Africa poses long-term economic, financial, and operational risks to investors through several developmental practices. These risks have raised some concerns regarding a lack of accountability, increased expenses, an unstable economy, the spread of economic crisis from one region to another, and constraints in enforcing the rule of law. The South African construction industry has been making progress in implementing risk management practice through different frameworks, and the result has been great over the years. However, there is still a lot of work to be done in terms of its full adoption, challenges, benefits, etc. It makes up the country's reception toward risk management practice.

Keywords Construction risk · Project delivery · Project risk · Risk influence · Risk practice · Risk types

Introduction

In trying to define risk and what it entails across several industries, the formulation and framework of its concepts bring about the difference in risk that is seen and addressed in the industries. For example, the stakeholders in the construction industry see risk as a way of removing impediments that will reduce the efficiency of a project from planning to completion. Furthermore, risk management in a project is defined as bringing value as a whole life cycle casualization in project delivery.

Risk is becoming an increasingly critical issue in the construction sector. While losses incurred during the cause of operation are normally high, there is the possibility of experiencing a catastrophic failure if risk is not addressed (Embrechts et al. 2003). More recently, there has been a significant shift in the insurance sector toward a better awareness of risk encountered by individual firms and the sector as a whole (Acharyya, 2012). The construction business is usually considered risky

A. E. Oke et al., *Risk Management Practices in Construction*,
https://doi.org/10.1007/978-3-031-35557-8_6

because of its complexity. The business involves a large number of stakeholders, and this implies that both internal and external determinants will pose significant risks. Notably, the construction industry has a low-risk analysis grade when compared to other industries (Laryea & Hughes, 2008).

However, one should bear in mind that no construction project is without risk. On the other hand, risk can be controlled, lessened, shared, transferred, or accepted. According to Project Management Institute (2008), risk is seen as a multifacet concept. Project risk management is commonly focused on identifying, analyzing, and responding to project risk. Risk from the perception of construction can be seen as the possibility of an event or series of events occurring during the construction process. This process consists of different phases; this might make it difficult to discern cause and effect, dependency, and correlations. As a result, these risks can have a major impact on decision-making and thus affects project performance.

Due to the nature of projects, which include so many interrelated and independent operations, several risk variables will affect the performance of such a project. Developing an effective risk management plan involves identifying and categorizing the main risk factors that impact the project's objectives the most.

In the study by Mark et al. (2004), the risk is the probability of experiencing complications and problems during the course of executing a project while trying to achieve its objectives. Therefore, construction firms must face unclear risks by analyzing their impacts on project objectives. Since the level of risk determines which project is riskier, preparing for such risk will give better condition toward mitigating them throughout the construction phases (Zayed et al. 2008). Several organizations have recently recognized the importance of proper risk management due to the obvious effects on project performance. Effective risk management is now an important factor to consider when evaluating general management measures in an organization, as its absence or presence can significantly impact the project's final outcome.

South African Construction Industry

The construction industry in South Africa plays a major role in the socioeconomic development of the country (Construction Industry Development Board (CIDB), 2012). Ofori (2007) and Giang and Pheng (2011) attested to this and further mentioned some of the ways the industry contributes to economy's growth. These include provision of employment (particularly for the least skilled people in society), contribution to the growth and spread of technology, creation of opportunities for businesses, and enabling improvement of the user's quality of life.

The country's population growth is a critical factor in maintaining the strength of the construction sector. In the report, according to the United Nations Population Fund (2022), the population of South Africa is expected to reach 72.9 million by 2050, with urban centers housing 62.1% of the country's current population. Therefore, the urbanization rate is expected to grow at an annual rate of 1.21%. As the country's urbanization and population are on the rise, housing needs and urban development are expected to be major growth drivers for the construction industry (Oxford Business Group, 2016).

Risk Management in the South African Construction Industry

The business environment in general, and in South Africa in particular is laddered with risks related circumstances. In their study, Visser and Joubert (2008) identified some of the major risks affecting the South African construction industry: lack of vital resources, poor project management and business risk management, exposure of the contract to risk during the tendering process, and government laws.

Risk assessment somehow in one way or the other affects project performance. There is a clear difference between risk and performance, and risk has a direct impact on the level of performance attained, while performance may be viewed as the measure of effectiveness and efficiency (Anthony, 1965). Efficiency as an included term can be referred to as the ratio of input to output consumption, and effectiveness expressed refers to the degree to which a planned outcome is achieved.

Risk has always been seen through the perspective of insurance, with little integration with other managerial functions. Global competition and market instability have brought risk management to the center of thought in the business world. It is essential to note that risks do not just relate to simply unexpected events; they can also refer to possible events that have an effect on the effectiveness and efficiency of operations, thereby influencing an organization's long-term performance.

On the other hand, the management aspect of the risk will normally concentrate on occurrences that could have major influences on the organization's performance or profile. Therefore, there is a need to assess the interconnectivity of risk since it impacts not only a particular activity but also a number of associated and seemingly unconnected activities. Ideally, all organizational activities are vulnerable to risk; however, other organizational procedures can mitigate the risk's impact (Ritchie & Brindley, 2007).

Prior to South Africa hosting the World Cup in 2010, some infrastructures were put in place to add glamor to the occasion. These infrastructures added more value to the already thriving South African construction industry. Chihuri and Pretorius (2010) assessed some of these infrastructures and identified some major risks encountered during the construction process. These risks include the following:

1. *Escalating Cost*: It is a result of unstable exchange rates and a steady increase in the cost of construction materials.
2. *Power Shortage*: It prompts concerned parties to seek alternative sources of power supply.
3. *Skill Shortage*: This compels the contractors to look for skills outside the country, thereby raising the project cost higher than anticipated.

The study also emphasized that risk management is still not widely practiced and that even in organizations where it is the principles are not fully implemented throughout the project life cycle. Further findings from the study showed that 71% of projects where risk management practices were fully implemented had a general success in project delivery. Serpella et al. (2014) also agree with this view with the claim that there will always be delays, cost overruns, and disputes as regards the contractual agreement if there is no efficient project management function to address the risks and uncertainties that comes with every project.

South Africa Risk Management Standard

The Standard Bureau of Standards (SABS), which serves as the national standardization institution, is governed by the Standards Act, 2008 (No. 8 of 2008). The institution has adopted ISO 31000:20:2018, and its risk management standard was reissued as SANS 31000: 2019. The ISO has released several standards that serve as risk management guidelines across different countries. In Borghesi and Gaudenzi (2013), this standard offers a consistent risk lexicon and terminology to encourage the exchange of information, the development of metrics and the communication of results.

The ISO 31000:20:2018 as presented by Sousa et al. (2012) addressed some important areas of risk management as stated as follows.

- It defines some key terms in risk management, such as risk, event, stakeholder, risk source, likelihood, control, consequence, and risk management itself.
- It provides principles for effective risk management (ERM) which basically creates and protects value. It also provides fundamental principles that should be considered when developing an ERM framework for an organization.
- It also addresses the risk management process, which involves risk identification, risk analysis, risk evaluation, risk treatment, monitoring, review and recording, and reporting.

Process/Framework for Risk Management in the South African Construction Industry

The control of risk and the management of risk outcomes are two fundamental procedures that characterize the risk management process. These basic processes consist of an array of activities that differs according to their complexity. The risk management process consists of the following subprocesses, as stated by (Bowden et al. 2001).

- *Risk Identification*: This particular process involves a well-detailed identification of risks that can have serious effects on the organization subprocess.
- *Risk Evaluation*: This is the most significant part of the risk management process, where the risk level is quantified by analyzing the rate at which an event might occur and the effect of such occurrence.
- *Risk Control*: This process involves the evaluation or design and implementation of risk treatment measures with the aim of controlling the effect of the risk. The management team is often responsible for the identification process, while the staffs are responsible for the execution.
- *Risk Monitoring*: This involves monitoring and evaluating the effectiveness of the risk treatment measures implemented by the organization.

Sources and Types of Risk

Risk identification aims to identify all forms of risks. Samson (1987) identified some of the possible sources of risk, which were broadly construed to include the following:

- The physical environment is in reference to the climatic condition.
- The social environment deals with human values.
- The political environment relates to the effects of monetary policies.
- The legal environment relates to irregularities that occur as a result of legal disparities.
- The operational environment concerns injuries suffered by employees during the manufacturing operation.
- The economic environment determines the economic variables such as interest rate and inflation.
- The cognitive environment has to do with perception against the reality.

Requirements for Effective Risk Management

There are some influencing factors that are significant in ensuring effective risk management. These factors are organizational processes, systems, and governance. One other important factor is the culture of an organization, which comprises the following elements (Page & Spira, 2004):

- Organizational operations should be based on well-informed, rational, and well-structured judgments.
- Long-term growth with little volatility should be sought.
- The order of priority in an organization should follow soundness, profitability, and growth.

In an organizational framework for assigning risk management, delegations should be designed and followed strictly by the concerned task performer. Additionally, thorough risk policies and processes for risk management should be designed and executed. Control over the processes should be performed by independent group supervision.

Recently, there has been a growing awareness of the importance of risk management; with this development, management should delegate this independent supervision responsibility to internal audit. However, good cooperate governance practices the deployment of a distinct risk management department for risk supervision. The internal audit's duty is to offer independent assurance to management and the board about the sufficiency of risk functions and the extent to which organizational objectives are met. Unexpected events that result in unfavorable risk situations would require more funds, time and resources from the organization, and risk the company's reputation. In contrast, the lack of such issues allows companies to use their resources better.

Challenges of Risk Management Practices in South Africa

Chihuri and Pretorius (2010) identified some key factors that impede the implementation of risk management throughout the life cycle of construction projects in South Africa.

1. *Lack of Appreciation*: There is a lack of appreciation for the enormous benefits that a structured risk management strategy can provide a project. Quite a number of engineers and project managers in South Africa are still uncertain about how risk management would contribute to overall project success.
2. *Time*: The risk management techniques require time for their implementation; thus, it affects the construction period of the project.
3. *Knowledge*: There is a general lack of understanding as to what the risk management process involves. There is a need for construction professionals to be trained so as to understand the benefits of this process.
4. *Perception*: There is this general perception of the risk management process involving a whole lot of cost.

Benefits of Risk Management in the South African Construction Industry

Risk management involves the use of the right tools to address the issue of risks. If all risk management processes are followed correctly, they can offer numerous benefits to engineering and construction projects in general. Shunmugam and

Rwelamila (2014) identified some of the major benefits of implementing risk management in South Africa.

- It allows for better decision-making.
- It allows for planning and prioritizing skills.
- It allows for well-organized resources and financial allocation.
- It allows for the prediction of problems, allowing for the best use of resources to combat and avoid detrimental outcomes.
- It improves the likelihood of completing the business plan within the stipulated time frame and cost.

Conclusion

The South African construction industry can improve its already exciting infrastructural development by inculcating functional management practices. One of these practices is risk management, introduced from the onset of construction in order to detect, plan, and mitigate failures (predicted and perceived) right from project reconnaissance to completion. Even with their exposure, the South African industry still needs to inculcate risk management process more to continue the rise to the summit in infrastructural development. This will aid all the parties concerned in project actualization and, subsequently, the country's economy development.

References

Acharyya, M. (2012). *The scope of developing optimization models for insurer's operational risk from risk-return trade-off perspective: Society of actuaries*. Retrieved from http://www.soa.org/Files/Research/Projects/The-Scope-of-Developing-Optimization-Models-for-Insurer-s-Operational-Risk-from-Risk-Return-Trade-Off-Perspective.pdf

Anthony, R. N. (1965). *Planning and control systems: A framework for analysis*. Harvard University School of Business Administration, Division of Research.

Borghesi, A., & Gaudenzi, B. (2013). *Risk management: How to assess, transfer and communicate critical risks*. Retrieved from https://link.springer.com

Bowden, A. R., Lane, M. R., & Martin, J. H. (2001). *Triple bottom line risk management. Enhancing profit, environmental performance, and community benefits*. Wiley.

Chihuri, S., & Pretorius, L. (2010). Managing risk for success in a South African engineering and construction project environment. *South African Journal of Industrial Engineering, 21*(1), 63–77.

Construction Industry Development Board (CIDB) (2012). The construction industry as a vehicle for contractor development and transformation, .

Embrechts, P., Furrer, H., & Kaufmann, R. (2003). Quantifying regulatory capital for operational risk. *Derivatives Use, Trading & Regulation, 9*(3), 217–233.

Giang, D. T. H., & Pheng, L. S. (2011). Role of construction in economic development: Review of key concepts in the past 40 years. *Habitat International, 35*(1), 118–125.

Laryea, S., & Hughes, W. (2008). How contractors Price risk in bids: Theory and practice. *Construction Management and Economics, 26*(3), 911–924.

Mark, W., Cohen, P. E., & Glen, R. P. (2004). *Project risk identification and management.* Retrieved from https://www.proquest.com/openview

Ofori, G. (2007). Construction in developing countries. *Construction Management and Economics, 25*(1), 1–6.

Oxford Business Group. (2016). *Long-term growth forecast for South African construction sector remains positive.* Retrieved from https://www.oxfordbusinessgroup.com/overview/weathering-storm-infrastructure-and-residential-projects-give-cause-optimism

Page, M., & Spira, L. (2004). *The Turnbull report, internal control and risk management: The developing role of internal audit.* The Institute of Chartered Accountants of Scotland Publisher (ICAS).

Project Management Institute (PMI). (2008). *A guide to the Project Management Body of Knowledge (PMBOK guide).* Project Management Institute Incorporation.

Ritchie, B., & Brindley, C. (2007). Supply chain risk management and performance: A guiding framework for future development. *International Journal of Operations and Production Management, 27*(3), 303–322.

Samson, D. A. (1987). Corporate risk philosophy for improved risk management. *Journal of Business Research, 15*(2), 107–122.

Serpella, A.F., Ferrada, X., Howard, R., & Rubio, L. (2014). Risk management in construction projects: A knowledge-based approach. In *Procedia of social behavioural sciences, selected papers from the 27th IPMA (International Project Management Association), world congress*, Dubrovnik, Croatia, 2013 119, pp. 653–662. https://doi.org/10.1016/j.sbspro.2014.03.073.

Shunmugam, S., & Rwelamila, P. D. (2014). An evaluation of the risk management in South African construction projects. In *Proceedings of the Project Management South Africa (PMSA) conference,* September 29–1 October, Johannesburg.

Sousa, V., Almeida, N. M., & Dias, L. A. (2012). Risk management framework for the construction industry according to the ISO 31000:2009 standard. *Journal of Risk Analysis and Crisis Response, 2*(4), 261–275.

Standard Bureau. (2008). *Standards Act (Chapter 43:07) Number 8 of 2008.* Retrieved from https://www.dffe.gov.za

United Nations Population Fund. (2022). *World population dashboard-South Africa.* Retrieved from https://www.unfpa.org

Visser, K., & Joubert, P. (2008). *Risk assessment modelling for the South African construction industry.* Retrieved from https://www.researchgate.net

Zayed, T., Amer, M., & Pan, J. (2008). Assessing risk and uncertainty inherent in Chinese highway projects using AHP. *International Journal of Project Management, 26*(4), 408–419.

Chapter 7
Risk Management Practices in Sri Lanka

Abstract Risk management is considered an emerging concept in the Sri Lankan construction industry. It is gradually gaining prominence because of rising construction activity, increased complexity and sophistication in construction projects, and strong market competition. This chapter addressed risk concepts and identified the risks associated with construction projects. It also highlighted the risk management processes and the tools that can be employed to execute each phase. It was concluded that there was a certain constraint to the application of risk management in Sri Lanka, as most clients were not proficient about risk management benefits and felt that contractors and consultants have the obligation to bear the risk effects by their own cost. Also, clients that were willing to actualize risk management practices in their projects were not prepared to spend extra cost in implementing risk management process.

Keywords Construction risk · Project risk · Project success · Risk actualization · Risk practice · Risk management

Introduction

The construction industry is highly diversified and highly unpredictable in its nature. There are numerous notable categories of construction that are distinct from one another. They include houses, non-residential buildings, road, utilities, and modern construction (Keoki et al., 2008). Construction projects incorporate brand-new construction, remodeling, and demolition for both private and public projects. This is to meet the ever-increasing demands of the clients. In addition to these projects, open project works like lanes, streets, parkways, utility plants, scaffolds, tunnels, and bridges are also constituents of the construction industry.

The success factors for any project include early finish time, within the budget plan, and imperative execution (specialized prerequisite). The boundary to these accomplishments is the adjustment in the project condition. The issue duplicates with the measure of the projects as vulnerabilities result in increment with the

A. E. Oke et al., *Risk Management Practices in Construction*,
https://doi.org/10.1007/978-3-031-35557-8_7

estimate (Dey & Ogunlana, 2002). Vast construction projects are presented to an unverifiable situation due to factors such as arranging, planning and development unpredictability, nearness of different involving groups, (clients, specialists, contractual workers, providers, etc.), assets (labor, materials, machinery, and assets) accessibility, environmental factors, the financial and political condition, and statutory controls. Olsson (2007) asserted that there are vulnerabilities in routine daily existence, institutions, and projects, which pose an obvious threat to the business. These vulnerabilities come in risk that occurs through many factors either at the onset of construction or throughout the phase of the construction process.

It has been demonstrated by Hillson (2004) that there is a relationship between uncertainty and risk. Risk is an estimated uncertainty, while uncertainty is a risk that cannot be estimated or assessed. Wang et al. (2004) described risk as a multifaceted occurrence with a probability of a harmful outcome on project objectives.

Risk was also defined by Kartam and Kartam (2001) as the likelihood of some unpredictable, undeterminable, and unwanted outcomes causing the expectations of profitability on a specific investment to be altered. A risk may also provide an opportunity, but the reality that the majority of risks have bad outcomes has caused many people to exclusively examine its negative aspects (Hillson, 2011).

Over the previous decades, numerous projects have experienced high cost and/or schedule fluctuations, ultimately resulting in the failure of these projects. As a result, systematic risk management is required to allow for the early detection of these risks (Abdelgawad & Fayek, 2010). Despite the fact that risk management research in the construction industry has grown significantly in recent years (Forbes et al., 2008), construction projects are exposed to risk from the moment they are conceived (Schieg, 2006) and are believed to have higher inherent risk due to the engagement of numerous contracting parties such as clients, contractors, designers, and engineers (El-Sayegh, 2008).

Risk management is generally recognized as an essential component of project management (Olsson, 2007). Perhaps one of the most challenging tasks is defining which risks are relevant to the project and how they should be prioritized among other risks (Anderson, 2009). This is a critical procedure as most project managers are aware that effective risk management is critical to project success (Baloi & Price, 2003).

Risk management is the cognitive process of detecting and analyzing risk before taking steps to reduce it to an acceptable level (Tohidi, 2011). Furthermore, Project Management Institute (PMI) (2007) expatiated risk management in construction as a detailed and precise approach for identifying, investigating, and responding to risks to meet the project objectives. Implementing risk management at the earliest stage of a project is critical, especially when crucial decisions such as alignment and construction methods can still be modified (Eskesen et al., 2004).

The Sri Lankan construction industry is growing toward the trend of risk management. Even though its adoption is recent, efforts have been made to implement the practice as much as possible since the country's infrastructural development has improved when compared to previous years. Also, due to increased project complexity and sophistication in construction projects, along with fierce competition in the market, construction professionals identified risk management as a functional way of dealing with emerging possibilities. In order to arrive at a realistic bid price for a planned or already project, it is vital to have a sense of the importance of risks and how they are allocated and distributed within these phases of the construction process (Perera et al., 2009).

Concept of Risk, Uncertainty, and Opportunity

Concept of Risk

Risk is ever inherent while drawing conclusions based on presumptions, anticipations, and appraisals of the future. This word describes an event in which the actual outcome of a specific result or action is expected to diverge from the predicted value (Raftery, 1994). In its many forms, risk can be quantified in a variety of ways depending on how it is viewed across different industries.

Further expatiation on risk can be seen as a condition in which the outcome of an event is conceivably predictable based on statistical chance. This characteristic distinguished it from uncertainty. It simply implies that there is cognition near a risk as a result of a combination of variables. This is in contrast to uncertainty (Smith et al., 2006). Risk is frequently described in terms of its chances and consequences along with its impact on various targets. According to Loosemore et al. (2006), a predicted outcome must have a probability value between 0 and 1, revealing a probability distribution in which the outcome is either unacceptable or certain to occur. As a result, once a conclusion is drawn regarding a sequence of conceivable outcomes, and after known odds may be attached to predict those outcomes, the presence of risk is established (Smith et al., 2006).

Hillson and Murray-Webster (2005) highlighted an intriguing tendency discovered while analyzing various officially published risk management guidelines. According to the authors, prior to 1997, individuals had a negative sense of risk, and as a result, the risk was synonymous with threat, with the condition being equivalent to hazard, danger, etc. Although the concept of risk described in many risk management literature studies has evolved since 2000, the vast majority of official standards plainly see risk as more of a threat rather than an opportunity or even a combination of both. In addition, PMI (2000) defined risk as an unpredictable consequence or occurrence that when it occurs it has favorable or damaging effect on the project's objectives. Risk can also be defined as the exposure to the theory of financial loss or gain, physical harm or injury, or time lag as a result of the uncertainty involved in following a trend of activity (Chapman, 1991). Smith et al. (2006) concluded by stating that risk entails ranking of possible outcomes before reaching a conclusion; these possible outcomes require facts for it to be understood clearly by all the involved parties.

Concept (Definition) of Uncertainty

Uncertainty is simply the probability of an event occurring in which the chance distribution is undefined. It can also be defined as the occurrence of an event about which little is known except that it may occur (Smith et al., 2006). Hence, there is no necessity for a fact for a conclusion to be made at a fixed moment in time (Winch, 2010). As a result, when an activity produces more than one potential result, but the likelihood of each result is unknown, the occurrence of uncertainty occurs (Smith et al., 2006).

Opportunities and Threats

In order to provide effective risk management on a project, it is essential to understand the relationship that exists between opportunities and threats (Hillson, 2004). Risk is not necessarily defined as the probability of having entirely undesirable outcomes; rather, it should as well admit the possibility of beneficial outcomes (Smith et al., 2006). Therefore, there is a need to consider threats and opportunities before establishing any conclusion, which must then be dealt with appropriately and effectively. It is important to also take into consideration related opportunities and not just focus solely on reducing possible threats. At the same time, it is not suggested to chase after opportunities while not being alert to possible threats (Chapman & Ward, 2003).

Opportunities and threats necessitate a degree of uncertainty, which has the ability to have an impact on the targets. An opportunity is a set of terms or an unpredictable outcome that, if realized, would benefit the overall success of the project. Threat, on the other hand, may be characterized as an unpredictable outcome or precondition that, if realized, would be detrimental to the overall success of the project. Therefore, from these definitions, it is evident that the primary distinction between these two terms is the sort of result they obtained/influenced on the target. Drawing a parallel in their definitions, it appears logical to mix these two together under a general definition that combines the component of uncertainty with the capacity to impact objectives which is how risk is modernly described (Hillson, 2004).

The Risks Involved in the Construction Sector

Risk in the construction sector is broadly defined as an event that has the potential to negatively impact the outstanding objectives of a project in terms of cost, time schedule, and quality (Dai et al., 2009). Additionally, it is well known that the construction industry is far more exposed to risk and uncertainty than any other industry (Tah & Carr, 2001). This may be attributable to intrinsic construction sector traits such as high levels of complexity, a high level of engagement (Hwang et al., 2014), exposure to project environments, tight schedule (Klemetti, 2006), and enormous and massive nature of the project (Thompson & Perry, 1992).

The engagement of a varied range of stakeholders and parties at each stage of the project amplifies the effects of these traits (Dikmen et al., 2008). Project risks may have an impact on all phases of the project (Tadayon et al., 2012). As a result, the success of construction projects depends on these stakeholders' ability to effectively manage the risks associated with the projects at hand (Ren, 1994).

Project Risk Classification

Smith et al. (2006) classified project risks into three groups: known risks, known unknown risk, and unknown unknown risks. The distinction between these categories is the diminishing power to predict or anticipate the risks. It is the diminishing ability to predict or anticipate risks that distinguish these groups from one another. The categories are as follows.

1. *Known Risks*: The known risks category includes risks that have been identified by the project team. They frequently occur in all construction projects and are therefore unavoidable, thus resulting in modest variations in the cost of material and efficiency (Smith et al., 2006). This risk category can be described as the rational prerequisite of risk in which the risk source has been identified, and the chance of an event happening in the risk case has been assigned (Winch, 2010).
2. *Known Unknowns*: The known unknown risks category is fairly certain indicating that little is known about the chance of their occurrences or the implications of their occurrences (Smith et al., 2006). This risk category can be described as the rational prerequisite of uncertainty where the source of the risk has been identified and comprehended, at the very least.
3. *Unknown Unknown*: In this context, "unknown unknowns" refer to the cognitive precondition of uncertainty in which someone could have had personal experience with the risk sources and possibilities but chooses to keep that information hidden. Because the source of the risk is unknown, it is impossible to predict the consequence of the risk (Winch, 2010). In this way, risk outcomes are situations whose outcome and the likelihood of occurrence are unknown, even to the most experienced and professional members of a project team (Smith et al., 2006).

Stakeholders' Attitudes Toward Risk

Understanding the principles and practices of risk management require evaluating different attitudes toward risk (Baranoff & Kahane, 2009). Hence, in order to conduct a successful evaluation of decision-makers' and stakeholders' risk attitudes while managing risk in construction, a complete understanding of their risk attitudes is essential (Wang & Yuan, 2011). Even more so, given that the construction industry has always struggled with a lack of cognitive retention and communication skills, which particularly necessitates more reason for the suggestion action (Liu et al., 2007). Risk attitudes can come in the form of the following:

1. Risk-averse.
2. Risk-neutral.
3. Risk-seeking.

Contrasting attitudes toward risk exist among people, and the individualist unique attitude will impact how they grasp risk and react to risk (Raftery, 1994). Attitudes help increase an individual's dignity, and they communicate a person's sense of self-identity and conduct esteems to others. These perceptions are consequently crucial to managers because they govern how people behave in response to a given input and allow perceptivity to be converted into propelling mechanisms. Individuals' attitudes are based on their own personal favorable or unfavorable ratings, opinions, and cognitions about the implication of definite conduct (Teo & Loosemore, 2001).

As a result, people's risk attitudes are a reflection of their individual feelings and qualities and the managerial environment in which they live. In this way, Winch (2010) defines project managers' risk preferences as their tendency or desire for the degree of risk and uncertainty that they are ready to accept in the same decision state of affairs. Wang and Yuan (2011) explained how different project managers make disparate and sometimes even contradictory assessments in the same decision situation. The author explained further that the framework is built on the three contrasting attitudes previously articulated, and it recognizes a variety of decision-making measures that are categorized according to their risk visibilities.

Risk-Averse

As soon as people and groups become uncomfortable with uncertainty, they become risk-averse. The traits of this instance of attitude are more common, and they are signaled and confirmed by highly successful working practices. Irritation and increased sensitivity are caused by the presence of a threat. As a result, there is a tendency for assertive risk reactions that diminish the risks. A risk-averse attitude, on the other hand, may underestimate the magnitude of possible chances (Hillson & Murray-Webster, 2005). They hope to acquire property in order to get as much security as possible in order to alleviate their suffering (Baranoff & Kahane, 2009).

Risk-Neutral

People and organizations with a risk-neutral attitude are more likely to engage in initiatives that have a high likelihood of success. Hence, they consider contemporary risk-taking a cost justified in light of the potential advantages.

This attitude is characterized by being fearless in times of challenge and seeing potential in obscurity. This approach places a strong emphasis on seniority once it concerns threats and potential opportunities. Thus, simply employing a procedure that is projected to yield significant benefits (Hillson & Murray-Webster, 2005).

Risk-Seeking

People and organizations who personify a risk-seeking attitude are more likely to tolerate a somewhat unplanned move in the direction of the behavior of potential threats. On the risk technique, the risk-seeking individual tends to identify less threats due to the model in risk attentiveness. It is expected that threats will be underestimated in terms of their potential impact and likelihood of occurring. The risk-seeking attitude may exaggerate the significance of potential chances and engage in aggressive behavior when confronted with these opportunities.

An attitude can be defined in two ways; the first definition refers to the internal making of the human mind, and the second definition refers to the mental aspect of information with respect to the information. This second definition depicts lean management, which could be understood as an analogy for the inner access an organization or person gains to a given state of events. Some attitudes are entrenched, while others are more flexible; they all, however, provide an alternative making them situational reactions that can vary based on the work environment. If the impacts are both explicitly and implicitly described, the potential of modifying the attitudes is opened up. However, attitudes are not limited to non-inheritable characteristics of people or organizations; instead, they can be transformed, and this is important to the cause of discernment and handling risk attitudes (Hillson & Murray-Webster, 2005).

Sources of Risk in Construction

On a broad scale, the sources of risk associated with construction can be grouped into three categories as follows:

1. *Internal Risks*—risks that can be controlled (e.g., risks involving design, construction, management, and relationships).
2. *External Risks*—risks outside of one's control (e.g., financial, economic, political, legal, and environmental risks).

3. *Risks Associated with Force Majeure*—risks associated with the environment (weather condition), e.g., frost, typhoon, and hail.

Risk Management in the Construction Industry

The term "risk management" basically refers to the process of identifying, describing, and responding to project-related risks. It must maximize the likelihood and impact of positive outcomes while simultaneously reducing the likelihood and impact of negative outcomes to ensure the overall success of the project (Tipili & Ibrahim, 2015). Risk management can be considered a decision-making procedure, as it involves making a thorough assessment of a known risk and/or taking crucial steps toward mitigating the impacts and probabilities of the outcome of such risks. This is done to minimize the complexity of the risks and raise the chances of success (Goh & Abdul-Rahman, 2013).

Goals and Aim of Risk Management

A significant improvement in the construction project management performance can be achieved by implementing a risk management process. Rather than completely eliminating all project risks, the risk management approach seeks to reduce them to the greatest extent practicable. Its aim is to establish a coordinated structure that encourages project management to deal with project risks, particularly the most critical ones more effectively and efficiently than now exists (Wang et al., 2004).

Sri Lankan Construction Industry

The current growth of the Sri Lankan construction industry is primarily attributable to economic factors. The sector has benefited the most from the government's recent push to provide infrastructure and guarantee massive development projects are completed following the end of the country's ethnic war in 2009. In Sri Lanka, there have been developments in numerous infrastructure areas, including luxury residential projects, road projects, affordable housing, water projects, and port projects. These initiatives have substantially impacted the Sri Lankan economy (Investment Information and Credit Rating Agency (ICRA Lanka), 2011). According to the Central Bank Report (2016), construction operations in Sri Lanka have increased significantly in value from a prior increment of 2.7–14.9%, accounting for 7.6% of the country's Gross Domestic Product (GDP).

However, there are some persistent challenges in the Sri Lankan construction industry which are comparable to those encountered in every developing country around the world. Some of these challenges are high construction material costs, scarcity of high-quality steel, scarcity of skilled labor, a deficit of funds, pollution, and regulatory changes (ICRA, 2011).

Risk Management in Sri Lankan Construction Industry

Since only a few studies on risk management have been undertaken in Sri Lanka (Perera, 2006), there is not much to add to the existing body of knowledge about the island nation that is completely different from what have been discussed across this chapter so far. The majority of the risk-related research available on Sri Lanka construction industry is focused on the context of road construction, with few (or very few) studies fitting the criteria for conducting an effective risk investigation on construction projects (Perera et al., 2009).

Risk Management on Sri Lankan Road Project

In response to the ever-increasing volume of traffic on Sri Lanka's roads, the Road Development Authority (RDA) developed plans to construct a nationwide highway network in the future (RDA, 2006). Many factors, including the availability of interested parties, available resources, the physical, economic, and political environments, and legal and regulatory requirements, frequently accompany road construction projects. Wang and Chou (2003) pointed out that such risks have a major impact on the project's outcome. Since then, it became necessary to develop fundamental physical infrastructure in order to support the major types of real estate investment, particularly those involving roads, because the success or failure of such investments will have a long-lasting effect. Kitazume et al. (2005) described road projects as "social capital development projects that are exposed to a variety of risks throughout their life cycle." The lengthy construction and maintenance periods, as well as the broad geographic coverage, are the reasons behind this. With a wider time scale comes a greater likelihood of interference or external events that will have an impact on the project.

Developing countries' construction projects are vulnerable to unpredictable circumstances (Ebrahimnejad et al., 2010). Implementing risk management procedures will help avoid a wide range of project challenges (Tadayon et al., 2012). However, little information is available on the country's application of risk management practices except on road projects discussed. This can be attributed to a lack of focus on the risk management approach (Tadayon et al., 2012).

The sources of risks lined with road development in Sri Lanka are considered to be unique because of the country's emerging economy and geographical location.

There is evidence that contractors and clients do not pay close attention to such risks in this type of economy, resulting in cost time overruns and poor workmanship (Tadayon et al., 2012). Numerous Sri Lankan roads have been rebuilt or rehabilitated in recent years primarily due to degradation and increased traffic to the point where expanding or realigned certain sections were required. This thus necessitates identifying and properly allocating risk to contractual parties to reduce foreseen or unforeseen circumstances regarding the project.

Risk Identification on Sri Lankan Road Projects

In the study by Perera et al. (2009), the sources of risk in carrying out road projects in Sri Lanka were identified using some foreign-funded road projects that were close to completion. In order to minimize the complexity that can occur when analyzing several types of road projects at the same time, projects that use traditional procurement methods with ad measurement were chosen for consideration due to the fact that they are the most commonly used procurement method in Sri Lanka.

These projects were chosen because they required a project completion time of at least twenty-four (24) months since researchers believe that gathering risk-related data requires a longer period of time. Some risk sources were identified through a review of literature studies and concepts that were developed based on interview transcripts. Furthermore, the result of the research reveals that the risk sources were relevant and were then divided into four types of risk sources stated below.

1. *Technical and Contractual Risks*
   - Low value of preliminaries bill.
   - Necessary changes imposed by the supervising engineer.
   - Delay in handing over of project.
   - Detailed drawings.
   - Change in the scope of the work.
2. *Economic, Financial, and Political Risks*
   - Late payments.
   - Relying on foreign loans for the funding of the project.
   - Regulations and difficulty in obtaining clearance permits.
   - Inflation.
   - Change in policies and legislation.
3. *Managerial Risks*
   - Competent level of the contractor.
   - Dealing with utility agencies.
   - Regulations and difficulty in obtaining clearance and permits.
   - Inflation.
   - Change in policies and legislation.

4. *External and Site Condition Risks*

   - Uncontrollable natural forces.
   - Harsh weather conditions.
   - Unforeseen site ground conditions.
   - Inflation.

Risk Identification Techniques Used on Road Projects

Perera et al. (2009) identified some of the risk identification tools that were used on the researched road projects; they include the following:

- *Checklists*: Identification of potential risks was based on the information checklist documented from previous abandoned and failed projects in Sri Lanka to enable project managers to know the risks involved on the road project and ways of minimizing them.
- *Interviews with Experts*: Experts were engaged to seek out ideas on risk containment based on their experiences on project evaluations. This was done in order to obtain profound practical solutions to issues not on the preidentified list and also for receiving feedback on the risks associated with identified projects.
- *Past Experience*: This analyzes historical data on similar risks from previous projects as the basis for a breakdown for evaluation. Risks that might be prevalent in each phase of the project life cycle and for best risk reduction from successful projects were identified and assessed carefully.

Risk Analysis and Evaluation on Sri Lankan Road Projects

After reviewing several literature studies on risk management in Sri Lanka, it can be outlined that risk analysis assessments and evaluation of the identified risk sources which were pertinent to the Sri Lanka Road project were based on two different approaches as stated below:

The Triangulation Approach

This approach is the combination of two data analytical processes together to achieve a particular aim. Love et al. (2002) opinionated that this approach is advantageous since it allows both qualitative and quantitative data in the process of generalizing the findings. Also, more grounds can be followed, and relevant information can be obtained more across the respondents.

In their study, Love et al. (2002) identified multiple sources of risk evidence. These include semi-structured interviews with project participants, documents such as letters, weather records, bills of quantities, claims, non-conformity reports, variation orders, the program of work, public complaint reports, and interim valuation and progress reports. They also included archival records like historical meteorological data.

In a research conducted by Perera et al. (2014) on *enhancing the effectiveness of risk management practices in Sri Lankan Road construction projects, the Delphi approach*, analytic hierarchy process (AHP) method, and sensitivity analysis were used for risk analysis assessments and evaluation of prospective identified risk sources pertinent to Sri Lanka Road project.

- *Quantitative Approach*: Sensitivity analysis was used to determine the project's uncertain components that would have the greatest impact on the final outcome.
- *Qualitative Approach*: Delphi approach and analytic hierarchy process (AHP) method were also used. The mentioned practices are explained as follows.

Delphi Approach

This procedure involves identifying qualified experts, preparing acceptable questions for them to answer, and analyzing their responses (Outhred, 2001). According to Xia and Gu (2012), even if expert collective judgments are constituted of subjective views, they are more reliable than individual assertions, hence resulting in more objective conclusions being made. The Delphi approach ensures that the results of a project are based on the consensus of the experts who are working on a similar project.

Analytic Hierarchy Process (AHP) Method

This method is based on perceptions and is a multicriteria decision analysis method developed by Saaty in 1980. This process is used for eliciting decision-makers' preferences in the form of ratios using pairwise comparison matrices based on perceptions for ranking risk and resource allocation (Deb, 2012). However, this approach was selected for the purpose of allocating risks among the major stakeholders (clients/consultants, contractors, and engineers) to cover more research ground and gather enough information about the intended projects.

Risk Allocation on Sri Lankan Road Project

The risk-handling methods used for the Sri Lanka Road project were as follows:

1. The allocation of risks among contracting parties for all actual risk handling.
2. The allocation of risks through contract clauses.

Allocation of Risks Between Contracting Parties

1. *Employer*: Risks allocated to this party are risks related to uncontrollable natural forces, harsh weather conditions, necessary changes imposed by the supervising engineer, dealing with utility agencies, defects in design, delay in making payments, relying on foreign loans for the funding of the project, inflation, delay in handing over of project, change in policies and legislations, interaction or relation with the surrounding community, security and safety of the public, regulations and difficulty in obtaining clearance permits, changes in the scope of the work, and unforeseen site ground conditions.
2. *Contractor*: Risks allocated to this party are risks related to harsh weather conditions, necessary changes imposed by the supervising engineer, competent level of the contractor, dealing with utility agencies, defects in construction, defects in design, incorrect estimation of the work, inflation, low value of preliminaries bill, the inadequacy of labor and equipment output, interaction or relation with the surrounding community, procurement of needed resources, security and safety of the public, regulations and difficulty in obtaining clearance permits, detailed drawings, and unforeseen site ground conditions.
3. *Engineer*: This party is responsible for the risk of late approval.

Allocation of Risks Through Contract Clauses

1. *Employer*: Risks allocated to the employer by the contract clauses include risks related to uncontrollable natural forces, harsh weather conditions, necessary changes imposed by the supervising engineer, dealing with utility agencies, defects in design, delay in making payments, relying on foreign loans for the funding of the project, inflation, late approvals, delay in handing over of project, legislative changes, interaction or relation with the surrounding community, change in policies and legislations, relations with neighborhood, security and safety of the public, building regulations and difficulty in obtaining clearance permits, change in the scope of the work, detailed drawings, and unforeseen site ground conditions.
2. *Contractor*: Risks allocated to the contractor by the contract clauses include risks related to harsh weather conditions, necessary changes imposed by the

supervising engineer, competent level of the contractor, dealing with utility agencies, defects in construction, defects in design, incorrect estimation of the work, inflation, low value of preliminaries bill, the inadequacy of labor and equipment output, interaction or relation with the surrounding community, procurement of needed resources, security and safety of the public, regulations and difficulty in obtaining clearance permits, detailed drawings, and unforeseen site ground conditions.

3. *Engineer*: Risks allocated to the engineer by the contract clauses include risks related to defects in design, security and safety of the public, detailed drawings, and unforeseen site ground conditions.

Risk Allocation Findings on Sri Lankan Road Projects

The Federation Internationale Des Ingenieurs-Counseils (FIDIC) published a set of contract conditions (FIDIC, 1987), which was used in the selected projects. Given that risks are shared among contractual parties via contract provisions, the administration of construction risks was first examined using the conditions of contract (CC) used in the chosen cases. Since the conditions of the contract in both situations were the same, the analysis could be conducted on the same set of facts in the selected cases.

After that, the actual allocation, which is the allocation of risks that are not explicitly stated in the contract clauses but are agreed upon by the contractual parties, is identified, and the risks associated with these allocations are evaluated.

However, with the exception of a few contract clause modifications, it was discovered that for the vast majority of the risk factors identified, actual risk allocation and risk allocation through contract clauses were nearly identical. Contract clauses modifications that are used in place of actual risk allocations include the following:

The risk of uncontrollable natural forces that becomes a risk to the Employer was exposed through the sub-clause 20.4 (h) [Employer's Risks] that the Contractor must also bear some of the risk connected with this.

In accordance with the sub-clause 42.2 [Failure to Give Possession], the risk of Delay in the handing-over of project had been allotted to the Employer; nevertheless, the Contractor was also required to bear this risk as a result of irrecoverable difficulties that he had to encounter.

Under the sub-clause 52.3 [Variations Exceeding 15 percent], the Employer is responsible for the risk of Change in the scope of work in both projects, while the Contractor is also responsible for a portion of the risk associated with delay in handing-over of project.

In the three cases highlighted, the risks had been shared only with the employer by actual risk allocation, while according to contract clauses, both parties (employer and contractor) should bear the risk.

Model for Risk Management in Sri Lankan Construction Industry

Wang et al. (2004) defined the steps in the methodologies of risk management in the construction industry as follows:

- Risk planning.
- Risk identification.
- Risk analysis/assessments and evaluation.
- Risk response; risk monitoring and control.
- Feedback.

Risk Planning

Risk planning requires projecting how to advance and execute risk management procedures needed to guarantee that the point, type, and profile of risk management are proportionate to the scale of the risk, as well as its relative importance in relation to the project's overall importance. During the risk planning stage, the objectives of the projects are set out, and responsibilities are delegated to the project's key players (Project Management Institute PMI, 2004).

Risk Identification

Risk identification is the process of identifying and retaining records of the risks that are associated with a given situation. It can as well be described as the technique of analytically and perpetually identifying, evaluating, and categorizing the initial impact of the risks related to construction projects (Al-Bahar & Crandall, 1990) and the interdependency of these risks (Liu et al., 2016). The concept of risk identification appears to be straightforward and easily applied (Hassanein & Afify, 2007). It is an important metric since the risk management and risk analysis approach can only be applied to risks that have been identified as a possibility (Wang et al., 2004). When not carried out in time and properly, it has the potential to have negative implications on the growth and success of a project (Crnković & Vukomanović, 2016). It is possible that failure to identify potential risks will result in a deficiency throughout the process. This, in turn, might have a major impact on the organization's available resources.

Benefits of Risk Identification

Rostami (2016) highlighted some benefits of risk identification as follows:

1. It helps to determine the optimal and the most critical input data to use.
2. It helps to have a deeper understanding of the process significantly.
3. It helps in identifying potential risks and its possible impact.
4. It provides information to people (stakeholders) who are in charge of making decisions.

Risk Identification Categories

In accordance with Zou et al. (2007), risk categorization is an inherent component of risk identification. Risks of various types are classified into several classes based on the features that have been predetermined for them. Generally, risk can be classified as follows:

- *External*: These are politically related, regulation-related, and environmental- and market-related.
- *Internal*: These are risks that have to do with safety, customer satisfaction, cost, time, and quality, coupled with service-related.
- *Technical*: These are risks that have to do with the managerial, resource, productivity, design, client, and payment.
- *Unforeseeable*: These are risks that have to do with natural hazards.

Tools and Techniques Used for Risk Identification

The process of risk identification can be completed with the assistance of a variety of tools and procedures (Rostami, 2016). Brainstorming, the Delphi approach, interviews, cause and effect analysis, SWOT analysis, and presumption analysis are some of the tools and methodologies available. In Crnković and Vukomanović (2016), the first four procedures are fundamental techniques; however, the final two techniques are utilized in particular to enquire into a broader scope of prospective scenarios.

The four strategies identified below are commonly employed to detect risks in developing nations in the area of infrastructure (Tipili & Ibrahim, 2015).

1. *Checklists*: It helps identify possible events that might have caused failure in past projects, which can be useful in identifying risk. It also enables the project manager to understand the risk involved in a project and prompts their involvement in the whole process of risk identification. This will enable a more favorable reception of any tools or methods deployed to mitigate the risks.

2. *Interviews with Experts*: This involves seeking out ideas on risk containment based on experts' experiences. It also involves analyzing previously documented data for works that are identical and assessing identical past or present works. It also involves project reviews through lessons learned from experts to obtain feedback on project risks.
3. *Past Experience*: This involves reviewing data of similar projects carried out in the past to ensure risk reduction. This particular technique is limited in its usage as it cannot be applied to all projects.
4. *Brainstorming*: This approach is more suitable for projects that involve new risk, new management systems, or the development of preliminary checklists. This could be beneficial in risk management workshops.

Risk Analysis/Assessment and Evaluation

After risks have been identified, the next step is to ascertain their implications as entirely as conceivable in order to prioritize them ahead of the response management stage (Schieg, 2006).

Risk analysis is considered the process of requiring the vital rating of prospective risks, ordering those risks according to importance, and selecting the most critical risks by the management team (El-Sayegh & Mansour, 2015). When it comes to risk management, risk analysis is by far the most challenging process to master. This is because it requires determining the likelihood of a risk occurring and its influence on the project's objectives (Thomas, 2006). Its primary goal is to analyze risk by categorizing unavoidable occurrences, estimating the possibility that an unpleasant event will occur, and estimating the degree of such situations (Karimi et al., 2011). It is implied that it is the intermediate phase between risk identification and management. It incorporates uncertainty in both qualitative and quantitative forms in order to assess risk outcomes. The rating should focus mostly on risks with a higher probability of occurring or having negative effects (Wang et al., 2004).

Risk Analysis Approaches

There are two main approaches to risk analysis that are commonly employed. They are qualitative risk analysis and quantitative risk analysis, and semi-quantitative risk analysis as a subcategory (Choudhry & Iqbal, 2012).

- *Quantitative Approach*

It is predicated mainly on the chance dissemination of risks. However, if sufficient information is made available, it can provide objective solutions to problems. Furthermore, quantitative risk analysis represents risks as numerical figures, allowing them to be measured in terms of how well they perform in quality, time, and cost (Morledge et al., 2006). Risk assessment is accomplished by quantitative analysis, which determines the impact of each risk on a scale from high to low and the likelihood that it will occur (Zou et al., 2007).

Quantitative Analysis Techniques

The following are the fundamental quantitative techniques:

These are decision tree analysis; expected monetary value; fault tree analysis (FTA); fuzzy logic; probability distributions; and sensitivity analysis/tornado diagram (PMI, 2013).

1. *Decision Tree Analysis*: It is a valuable technique for drawing up the problem and assesses several alternatives. This technique is in the form of a tree diagram, which gives more illustration of the project, and its decisions will determine the consequences of these decisions (Mhetre et al., 2016).
2. *Expected Monetary Value Analysis*: This considers the odds component of the arrangement states and is established on a gain matrix.
3. *Fault Tree Analysis (FTA)*: It describes vital risk cases but measures one outcome in a specific time.
4. *Fuzzy Logic*: It is valuable wherever probabilistic information is abstracted. It is not as accurate as probabilistic techniques.
5. *Probability Distributions*: This is a technique employed to reveal the potential influence of a contrastive degree of uncertainty on project goals. Specifically, it assesses the impact of risks on the project's cost and timeline, estimating each task with three points, including the most speculative event scenario, the most likely scenario, and the worst-case scenario. The Monte Carlo simulation approach is almost always utilized for this type of analysis.
6. *Sensitivity Analysis/Tornado Diagram*: It is used to describe uncertain elements in a project that will have the greatest impact on its outcome. It is expressed as an order of magnitude. The goal here is to consider the sensitivity of varied risk framework factors on project outcomes by changing the values of one variable at a time and then showing the influence on the outcome.

Qualitative Risk Analysis Approach

When applying this approach, it is frequently necessary to rate the effects of risks and develop lists of risks from the smallest to the largest. This is done to facilitate an early breakdown of the identified risks (Zou et al., 2007). It is executed by a data-driven methodology (Banaitene & Banaitis, 2012). However, it is open to individualized perception, suspicion, and assessment. Thus, depending on the analyst's point of view, the results can vary significantly. Therefore, the qualitative technique continues to be the preferred strategy by the majority of professionals (Jarkas & Haupt, 2015). A risk-scoring matrix (also known as a probability/impact matrix) is a type of risk analysis tool that is commonly used in qualitative risk analysis.

Qualitative Risk Analysis Approach

The following are the fundamental qualitative techniques: brainstorming, expert judgment, causes and effect diagram, checklists, Delphi, event tree analysis (ETA), risk breakdown matrix (RBM), risk data quality assessment, probability and impact assessment, and risk matrix method are some of the most common qualitative analysis approaches.

1. *Probability and Impact Assessment*: This technique can be used to predict the chance of a given risk occurring in the future. When evaluating the impact of risk on the success of a project, it is crucial to include both chances and favorable outcomes, threats, and unfavorable outcomes. It is essential to establish the likelihood of a risk occurring in a project and the implications it may have on the project.
2. *The Risk Matrix Method*: This method, in conjunction with probability and impact, can be utilized to lay the groundwork for further evaluation. The priority score is computed by averaging the probability and impact. The priority score range, rate, and color are all provided to demonstrate the significance of each risk. Threats with a high priority score, indicating a high impact and possibility of occurrence, are regarded as high-risk and may demand an immediate response, whereas threats with a low priority score could be monitored more closely and given attention only if necessary.
3. *The Delphi Method*: When using this method, you will be tasked with selecting qualified experts, developing relevant questions to be presented to them, and analyzing their responses (Cabaniss, 2002).
4. *Expert Judgment*: This method is based on using a professional's opinion to determine the likelihood of failure and the likelihood of achieving the project's objectives.
5. *Risk Breakdown Matrix (RBM)*: Risk is classified into smaller units using a risk breakdown matrix.
6. *Risk Categorization*: This is used as a method of systematizing risks according to their sources, with the goal of identifying areas that are most vulnerable to certain risks. Due to how this approach is used, activities are broken down into little components and organized into a hierarchical series of activities. Additionally, the approach can include risk dependencies and prioritize them based on how quickly they require a response.

7. *Analytic Hierarchy Process (AHP)*: This method involves gathering information and using pairwise comparisons to rank the risks associated with various events. This method is limited to only a few pairwise comparisons.

Semi-Quantitative Risk Analysis Approach

This particular approach establishes a personal estimate of the oftenest with which risks occur as well as an objective assessment of risk outcomes (Mead, 2006).

Other Risk Analysis Techniques for Project Risk Management

Monte Carlo simulations and system dynamics are examples of computer-based simulation technologies that can be used in risk analysis for project risk management (Choudhry et al., 2014).

Monte Carlo Simulations This technique sums up the cumulative impacts resulting from a set of uncertain parameters. It is computationally expensive to use this technique.

Criteria for Selecting Risk Analysis

1. The nature and scope of the proposed project.
2. Readily available information.
3. The financial ramifications and the amount of time available.
4. The analysts' collective knowledge and experience.
5. The amount to which innovation has occurred.
6. The ultimate goal of the outcomes (Goh & Abdul-Rahman, 2013).

Risk Response

Risk response involves the development of alternatives and ascertaining activities to heighten chances while also reducing threats to achieving project objectives. This strategy necessitates the identification of parties who will assume responsibility for each harmonized risk response, and the effectiveness of this stage will determine whether the project risks will step up or drop off (PMI, 2000). This also addresses the design of a strategy to mitigate potential threats while increasing the likelihood of success (PMI, 2004). It is necessary to implement adequate risk mitigation measures after the risks associated with the project have been identified and investigated. These risk mitigation techniques are based largely on the nature of the risk and the potential effects of taking action. Its main purpose is to increase risk dominance, reduce the destructive risk impact, and get rid of the potential possible impact. The criterion gets a lot more efficient when there is further dominance of one mitigation assess on one risk (Wang et al., 2004).

Types of Risk Responses

The six distinct risk reactions are retention, reduction, control, sharing, transfer, and avoidance (Kerzner, 2003). Alternatives for responding to risk need to recognize and acknowledge the significance of the risk. They also need to be financially cost-effective and honest in terms of project timing and also acceptable to the other parties involved (Goh & Abdul-Rahman, 2013).

- *Risk Retention*: It demands acknowledging the existence of a specific risk situation and devising an intended measure to bear the degree of risk without employing any exceptional drives to contain it (Kerzner, 2003).
- *Risk Reduction*: It is employed to convey the probabilities and effects of the risk low beneath a threshold that is accepted (Loosemore et al., 2006).
- *Risk Sharing*: It is primarily incurred through a contractual process to acquire a feeling of corporate obligation among the parties to the project (Loosemore et al., 2006).
- *Risk Control*: It does not totally remove risks in the project; rather, it accepts methods to reduce existing risks (Kerzner, 2003).
- *Risk Avoidance:* It is an act of unwillingness to acknowledge the risk or activity undertaken to ensure that the risk is not active to occur (PMI, 2004).
- *Risk Transfer*: It changes and alterations, as well as ownership from one party to another third party, without affecting the overall level of risk or lowering the importance of the risk sources (Smith et al., 2006).

In this phase, accessible alternatives and processes are acquired in order to promote opportunities and mitigate threats to project objectives. Each contracting party appears to be prepared to take a certain degree of risk and be well aware of its personal apportion of risks and the conditions under which losses will occur (Perera et al., 2014). The implication of each party's consciousness of the risks assigned to them, as well as the sequential planning required to address such risks, is assertive and subsequently adds to the project's success (Zou et al., 2007).

Risk Response Strategy in Developing Countries

In developing countries, the risk response approach includes both preventive (avoidance) and corrective techniques (Kartam & Kartam, 2001).

1. *Preventive Risk Management Technique:* A risk is best managed when it is eliminated early enough in the planning stage of the project. Early consideration of risks prior to the start of a project and the development of an effective plan for dealing with them are examples of preventive management approaches that are employed during the planning stage to prevent or reduce a necessary but unavoidable risk.
2. *Remedial Risk Management Technique:* In order to mitigate the consequences of risks and, if possible, entirely eliminate them, remedial risk management processes must be used. This is because not all risks can be managed at the early stage of the project and certain risks will develop during the operation stage of the project. Therefore, the manner of remedying the situation becomes critical.

How Contractors Respond to Construction Risk

In accordance with Wang and Chou (2003), contractors typically employ three methods to transfer risk in construction projects, which are the following:

1. Insurance companies to insurance companies.
2. Modification of the terms of agreement of the contracts to benefit the client or other parties.
3. Delegation of tasks or subcontracting of work to other parties.

Risk Monitoring and Control Stage

Risk monitoring and control are important stages in the project life cycle because they ensure that the desired outcomes of risk response execution are achieved throughout the project life cycle. Regularly, risk management documentation is reviewed and updated, and the outcomes of risk monitoring and control can serve as a lesson for future decision-makers (Morledge et al., 2006). The performance of risk response is evaluated from time to time as the project progresses in order to correct any errors in the strategy that have been implemented and to realign the strategy with the project objectives.

Risk Monitoring and Control Techniques

Risk Register

A risk register is a risk database that is used as a crucial tool in risk management to keep track of the progress of the risk management process (Cooper et al., 2005). The design of the register is determined by the organization, the nature of the projects, and the persons involved. It is essential that the organizations draft their own register that is appropriate for them in terms of order of magnitude for it to be fully implemented as intended rather than adding another strain to an already demanding study schedule. The register should be incorporated with a database to make enrolling, storing, managing, and categorizing data more convenient (Flanagan et al., 2007). The risk register contains a complete list of all known risks, along with the outcomes of their analysis and associated activity plans. It also includes an assessment of the risk's current status in relation to the specific risk. The risk register should be updated and re-examined on a regular basis throughout the duration of the project lifecycle.

Uses of Risk Register

A risk register is an essential tool for tracking and adjusting progress on risk mitigation processes. It aids in the description of new risks and the termination of low-terminated risks. Also, adjustment of the evaluation of present risk, among other things, can be checked with the use of a risk register (Potts, 2008). Risks that are no more critical due to avoidance or that have previously been dealt with can be removed from the register along with the action plans that were associated with them. The condition of activity plans, as well as specific risks, should be systematically re-evaluated and revised (Cooper et al., 2005).

Schieg (2006) opined that it is necessary to include new additional risks, risk conditions, and the progression of the process. Existing risks must be recorded, as well as the overall degree of damage incurred as a result of those risks. Furthermore, the author asserted that an important component of risk monitoring (which occurs in the final phase) is the internal constraint system, which delegates the responsibility for monitoring early indicators to certain individuals or groups. A reporting and meeting agreement for the project, as well as the organization as a whole, must be included in the contract for this process to be effective.

Feedback

Receiving feedback is critical for reviewing the treatment plan. It may be required to return to the identification stage if new risks emerge or the nature of existing risks changes during the project execution.

Conclusion

After a definite examination of different written literature on risk management in Sri Lanka, it can be outlined that currently, formal risk management M practice is rare among the Sri Lankan construction industry except for its one-time usage on road projects. The track record on risk management evolution has shown the value of risk management application for enhancing effective coordination among project teams for proper risk allocation on the executed road projects toward the management of cost, quality, and time, which are key objectives of each and every project. Moreover, research indicated constraints in the implementation of risk management to some degree in Sri Lanka, as most clients are not proficient in risk management benefits and feel that contractors and consultants have an obligation to bear the risk effects at their own cost. There is a need for different risk management processes to be inculcated in the construction firms for medium to large-scale project.

References

Abdelgawad, M., & Fayek, A. R. (2010). Risk management in the construction industry using combined fuzzy FMEA and fuzzy AHP. *Journal of Construction Engineering and Management, 136*(9), 1028–1036.

Al-Bahar, J. F., & Crandall, K. C. (1990). Systematic risk management approach for construction projects. *Journal of Construction Engineering and Management, 116*(3), 533–546.

Anderson, S. (2009). *Risk identification and assessment*. Retrieved from https://www.PMI.org

Baloi, P., & Price, A. (2003). Modelling global risk factors affecting construction cost performance. *International Journal of Project Management, 21*(4), 261–269.

Banaitene, N., & Banaitis, A. (2012). *Risk management in construction projects*. Retrieved from https://www.intechopen.com/chapters/38973

Baranoff, E., & Kahane, Y. (2009). *Risk management for enterprises and individuals*. Saylor Foundation.

Cabaniss, K. (2002). *Computer-related technology use by counselors in the new millennium: A Delphi study*. Retrieved from http://jtc.columbusstate.edu/vol2_2/index.htm

Central Bank. (2016). *Central Bank annual report 2016*. Retrieved from; https://www.cbn.gov.ng/Out/2018/RSD/CBN%202016%20ANNUAL%20REPORT_WEB.pdf

Chapman, C. (1991). *Risk, in investment, procurement and performance in construction*. E & FN Spon Publisher.

Chapman, C., & Ward, S. (2003). *Project risk management: Processes, techniques and insights* (2nd ed.). Wiley.

Choudhry, R. M., & Iqbal, K. (2012). Identification of risk management system in construction industry in Pakistan. *Journal of Management in Engineering, 21*(1), 42–49.

Choudhry, R. M., Aslam, M. A., Hinze, J. W., & Arain, F. M. (2014). Cost and schedule risk analysis of bridge construction in Pakistan: Establishing risk guidelines. *Journal of Construction Engineering and Management, 140*(7), 0401–4020.

Cooper, D., Grey, S., Raymond, G., & Walker, P. (2005). *Project risk management guidelines: Managing risk in large projects and complex procurements*. Wiley.

Crnković, D., & Vukomanović, M. (2016). Comparison of trends in risk management theory and practices within the construction industry. *Electronic Journal of the Faculty of Civil Engineering (e-GFOS), 13*(1), 1–11.

Dai, J., Goodrum, P., Maloney, W., & Srinivasan, C. (2009). Latent structures of the factors affecting construction labour productivity. *Journal of Construction Engineering and Management, 135*(5), 397–406.

Deb, M. (2012). Evaluation of customer's mall preferences in India using fuzzy AHP approach. *Journal of Advances in Management Research, 9*(1), 29–44.

Dey, P. K., & Ogunlana, S. O. (2002). Risk based decision support system for effective implementation of projects. *International Journal of Risk Assessment and Management, 3*(4), 189–204.

Dikmen, I., Birgonul, M. T., Anac, C., Tah, J. H. M., & Aouad, G. (2008). Learning from risks: A tool for post project risk assessment. *Automation in Construction, 18*(1), 42–50.

Ebrahimnejad, S., Mousavi, S. M., & Seyrafianpour, H. (2010). Risk identification and assessment for build–operate–transfer projects: A fuzzy multi attribute decision making model. *Expert System with Applications, 37*(1), 575–586.

El-Sayegh, S. (2008). Risk management and allocation in the UAE construction industry. *International Journal of Project Management., 26*(4), 431–438.

El-Sayegh, S. M., & Mansour, M. H. (2015). Risk assessment and allocation in highway construction projects in the UAE. *Journal of Management in Engineering, 31*(6), 0401–5004.

Eskesen, S. D., Tengborg, P., Kampmann, J., & Veicherts, T. H. (2004). Guidelines for tunneling risk management. In *International tunneling association, working group no. 2. Tunnelling and Underground Space Technology, 19*(3), 217–237.

Federation Internationale Des Ingenieurs-Counseils (FIDIC). (1987). *Conditions of contract for works of civil engineering construction*. Retrieved from https://www.justice-academy.com

Flanagan, R., Jewell, C., & Johansson, J. (2007). *Riskhantering i praktiken Byggnadsekonomi, Institutionen för bygg- och miljöteknik, Centrum for management i byggsektorn (CMB)*. Chalmers Repro. Retrieved from https://www.cmb-chalmers.se/wp-content/uploads/2015/10/riskhantering_praktiken.pdf

Forbes, D. R., Smith, S. D., & Horner, R. M. W. (2008). *A comparison of techniques for identifying risks in sustainability assessment of housing*. In: Dainty, A (Ed.) Procs 24th Annual ARCOM Conference, 1–3 September 2008, Cardiff, UK, Association of Researchers in Construction Management, 1135–1144.

Goh, C. S., & Abdul-Rahman, H. (2013). The identification and management of major risks in the Malaysian construction industry. *Journal of Construction in Developing Countries, 18*(1), 19–32.

Hassanein, A. A., & Afify, H. M. (2007). A risk identification procedure for construction contracts – A case study of power station projects in Egypt. *Civil Engineering and Environmental System, 24*(1), 3–14.

Hillson, D. (2004). *Effective opportunity management for projects: Exploiting positive risk*. CRC Press.

Hillson, D. (2011). *Dealing with business uncertainty*. Retrieved from http://www.risk-doctor.com/briefings

Hillson, D., & Murray-Webster, R. (2005). *Understanding and managing risk attitude*. Gower Publishing Limited.

Hwang, B. G., Zhao, X., & Toh, L. P. (2014). Risk management in small construction projects in Singapore: Status, barriers and impact. *International Journal of Project Management, 32*(1), 116–124.

Investment Information and Credit Rating Agency (ICRA Lanka). (2011). *Industry report on Sri Lanka: Construction industry in Sri Lanka.* Retrieved from http://www.icralanka.com/

Jarkas, A. M., & Haupt, T. C. (2015). Major construction risk factors considered by general contractors in Qatar. *Journal of Engineering Design and Technology, 13*(1), 165–194.

Karimi, A., Mousavi, N., Mousavi, S. F., & Hosseini, S. (2011). Risk assessment model selection in construction industry. *Expert Systems with Application, 38*(8), 9105–9111.

Kartam, N., & Kartam, S. (2001). Risk and its management in the Kuwaiti construction industry: A contractors' perspective. *International Journal of Project Management, 19*(6), 325–335.

Keoki, S. S., Sears, G. A., & Clough, R. H. (2008). *Construction project management – A practical guide to field construction management* (5th ed.). Wiley.

Kerzner, H. (2003). *Project management: A system approach to planning, scheduling, and controlling* (8th ed.). Wiley.

Kitazume, K., Miyamoto, K., & Sato, Y. (2005). Quantitative risk analysis of road projects based on empirical data in Japan. *Journal of the Eastern Asia Society for Transportation Studies, 6*(1), 3971–3984.

Klemetti, A. (2006). *Risk management in construction project networks*. Helsinki University of Technology.

Liu, J., Li, B., Lin, B., & Nguyen, V. (2007). Key issues and challenges of risk management and Insurance in China's construction industry. *Industrial Management and Data Systems, 107*(3), 382–396.

Liu, J., Zhao, X., & Yan, P. (2016). Risk paths in international construction projects: Case study from Chinese contractors. *Journal of Construction Engineering and Management, 142*(6), 0501–6002.

Loosemore, M., Raftery, J., Reilly, C., & Higgon, D. (2006). *Risk management in projects* (2nd ed.). Taylor and Francis.

Love, P. E., Holt, G. D., Shen, L. Y., Li, H., & Irani, Z. (2002). Using systems dynamics to better understand change and rework in construction project management systems. *International Journal of Project Management, 20*(6), 425–436.

Mead, P. (2006). *Conducting an effective and accurate assessment of project risk.* Retrieved from http://www.rics.org/NR/rdonlyres/CFD5F10B196B4A1E84925E03EEDF24B6/0/ProjectRiskAssessment.doc

Mhetre, K., Konnur, B., & Landage, A. (2016). Risk management in construction industry. *International Journal of Engineering Research, 5*(1), 153–155.

Morledge, R., Smith, A., & Kashiwagi, D. T. (2006). *Risk in building procurement.* Blackwell.

Olsson, R. (2007). In search of opportunity management: Is the risk management process enough? *International Journal of Project Management, 25*(8), 745–752.

Outhred, G. (2001). The Delphi method: A demonstration of its use for specific research types. In *Proceeding of the RICS Foundation, Construction & Building Research conference*, 1 September, Glasgow.

Perera, K. B. (2006). *Simulation modelling for time and cost overrun in road rehabilitation projects in Sri Lanka* (Bachelor of sciences unpublished dissertation). University of Moratuwa.

Perera, J., Dhanasinghe, I., & Rameezdeen, R. (2009). Risk management in the road construction: The case of Sri Lanka. *International Journal of Strategic Property Management, 13*(2), 87–102.

Perera, B., Rameezdeen, R., Chileshe, N., & Hosseini, M. R. (2014). Enhancing the effectiveness of risk management practices in Sri Lankan road construction projects: A Delphi approach. *International Journal of Construction Management, 14*(1), 1–14.

Potts, K. (2008). *Construction cost management: Leaning from caste studies.* Taylor & Francis Group.

Project Management Institute (PMI). (2000). *A guide to the project management body of knowledge*. Project Management Institute. Upper Darby.

Project Management Institute (PMI). (2004). *A guide to the Project Management Body of Knowledge (PMBOK guide)*. Project Management Institute Incorporation.

Project Management Institute (PMI). (2007). *Construction extension to the PMBOK guide* (3rd ed.). Project Management Institute.

Project Management Institute (PMI). (2013). *A guide to the Project Management Body of Knowledge (PMBOK guide)*. Project Management Institute.

Raftery, J. (1994). *Risk analysis: In project management* (1st ed.). Routledge.

RDA (2006). *Highway Development Plan: Road Development Authority (RDA)*. Retrieve from http://www.rda.gov.lk/source/highway_de-velopment_plan.htm

Ren, H. (1994). Risk lifecycle and risk relationships on construction projects. *International Journal of Project Management, 12*(2), 68–74.

Rostami, A. (2016). Tools and techniques in risk identification: A research within SMEs in the UK construction industry. *Universal Journal of Management, 4*(4), 203–210.

Saaty, T. L. (1980). *The analytic hierarchy process*. McGraw-Hill.

Schieg, M. (2006). Risk management in construction project management. *Journal of Business Economics and Management, 7*(2), 77–83.

Smith, N. J., Merna, T., & Jobling, P. (2006). *Managing risk in construction projects* (2nd ed.). Blackwell Publishing.

Tadayon, M., Jaafar, M., & Nasri, E. (2012). An assessment of risk identification in large construction projects in Iran. *Journal of Construction in Developing Countries, 17*(1), 57–69.

Tah, J., & Carr, V. (2001). towards a framework for project risk knowledge management in the construction supply chain. *Advances in Engineering Software, 32*(10), 835–846.

Teo, M., & Loosemore, M. (2001). A theory of waste behavior in the construction industry. *Construction Management and Economics, 19*(7), 741–751.

Thomas, D. R. (2006). A general inductive approach for analyzing qualitative evaluation data. *American Journal Evaluation, 27*(2), 237–246.

Thompson, P., & Perry, J. G. (1992). *Engineering construction risks: A guide to project risk analysis and assessment implications for project clients and project managers*. Thomas Telford.

Tipili, L.G., & Ibrahim, Y. (2015). Identification and assessment of key risk factors affecting public construction projects in Nigeria: Stakeholders' perspectives. In *Proceedings of the second Nigerian Institute of Quantity Surveyors Research conference,* 1–2 September. Federal University of Technology, Akure.

Tohidi, H. (2011). The role of risk management in IT systems of organizations. *Procedia Computer Science Journal, 3*(1), 881–887.

Wang, M. T., & Chou, H. Y. (2003). Risk allocation and risk handling of highway projects in Taiwan. *Journal of Management in Engineering, 19*(2), 60–68.

Wang, J., & Yuan, H. (2011). Factors affecting contractors' risk attitudes in construction projects: Case study from China. *International Journal of Project Management, 29(2)*, 209–219.

Wang, S. Q., Dulaimi, M. F., & Aguria, M. Y. (2004). Risk management framework for construction projects in developing countries. *Construction Management and Economics, 22*(3), 237–252.

Winch, G. (2010). *Managing construction projects* (2nd ed.). Wiley Blackwell.

Xia, Y., & Gu, X. (2012). The risk management of the intellectual property investment and financing of the high-tech SMEs. *Science of Science and Management of Science and Technology, 9*(2), 98–104.

Zou, P. X., Zhang, G., & Wang, J. (2007). Understanding the key risks in construction projects in China. *International Journal of Project Management, 25*(6), 601–614.

Chapter 8
Risk Management Practices in Tanzania

Abstract The economic importance of the construction industry in developing countries cannot be overstated due to the critical need for infrastructure development and housing. The Tanzanian construction industry accounts massively for the country's economic development. This chapter gave an insight into what risk management practice entails in general. Also, it discussed the practice in the said country along with challenges, impediments, and other related subsections. The study concluded that there is a low level of adoption of risk management practices among the construction professionals of the country. The study stated that the reason for this is the fact that risk management is still in its early stages. The chapter recommended that training and further awareness should be encouraged to increase the level of adoption and implementation in projects.

Keywords Construction industry · Project risk · Risk management · Risk mitigation · Risk practice · Sustainable construction

Introduction

The construction industry has always been a significant contributor to the growth of the economy of most nations. Lema (2008) stated that the Tanzanian construction industry contributes approximately 7% of the country's gross domestic product (GDP). Further study by Kikwasi (2011) noted that the construction industry accounts for about 9% of job creation and around 57% of capital development. Even though the sector is plagued with different challenges in cost and time overruns among all, the sector remains a cogent part of the country's economic development.

A risk is defined as the possibility or tendency of an event or problem occurring in relation to the success or achieving the desired aim of the project or unpredictable future occurrence, which would have impacts on the project's objectives (cost, quality, time, standard, etc.) (Mark et al., 2004). Furthermore, even as the effect of such occurrences is unpredictable, it has been hypothesized that risks can be mitigated to reduce their negative impacts on project success even though their

A. E. Oke et al., *Risk Management Practices in Construction*,
https://doi.org/10.1007/978-3-031-35557-8_8

occurrence is unavoidable during a project's life cycle (Chia, 2006). There are various sources of risk, and these include intrinsic uncertainties, institutional inadequacies, and company-related issues such as technological implementation, economic policy, weather and climatic condition, inflation, market competition, and bidding process (Karimi et al., 2011).

The practice of project risk management has been in existence since the mid-1980s and is considered to be one of the nine primary speciality areas of the Project Management Institute (PMI) body of knowledge (Tuysuz & Kahreman, 2006). The effects of risk can be lessened by observing, monitoring, managing, analyzing, and applying managerial skills and resources through synchronized socioeconomic and political stimulation. This is done to minimize the likelihood of unforeseen events occurring and to achieve the target for the project (Douglas, 2009). Effective and efficient risk management practices would lead to enormous benefits with respect to the success of a project, and such benefits include identification of favorable alternative course of action, adequate site investigation, proper project planning and budgeting, adequate price forecast, proper feasibility study, improved technological implementation, and proper and adequate estimates (Bannerman, 2008).

The ideology behind achieving sustainable construction is expanding, and higher standards are expected. This has led to the development of technology, improved material, and advanced technique applications. It is common knowledge that every new technology has its constraints and risks; the desire to achieve expected results through utilization of these modern technological techniques and materials has made the importance of project risk management more profound. Infrastructural development in any country would certainly enhance the economic growth of such a country and generate job avenues. The cost of undertaking such projects is enormous, and in order to deliver on time and standard, project risk management practice must be involved so as to prevent huge monetary losses (Bannerman, 2008).

Consequently, the risk is critical to the successful execution of a project within the stated time frame and budget. The construction industry is sometimes regarded as a high-risk venture because of its intricacy and strategic nature. It involves a vast number of parties and internal and external influences, and all this implies significant risks. When compared to other industries, the construction sector has a bad reputation for risk analysis and management (Laryea & Hughes, 2008). There are risks in all construction projects, but they can be managed, decreased or shared, transferred, or accepted; they cannot, however, be ignored (Latham, 1994). According to the experts, risk is a multi-facet concept, while project risk management involves identifying, analyzing, and responding to project risks (Project Management Institute, 2008). Risk management in the construction industry refers to the likelihood that a specific event or series of events will occur during the course of construction. There are several factors involved in the construction process, and it is frequently difficult to discern cause and effect, dependence, and correlations.

General History of Risk Management in the Construction

In the 1990s, risk identification and assessment of construction projects became the center of concentration in the industry. Different contractors developed various methods to analyze and assess these risks. The early techniques used to resolve risk-related issues in construction involved a systematic approach at the pre-contract stage to prevent time and cost overruns. This method required the identification of likely risk factors and assessing and proffering solutions to minimize their effects on the project. Risks and uncertainties (used interchangeably) in cost and time can be managed methodically throughout the estimation stage to mitigate any possible impacts (Birnie & Yates, 1991).

Chapman (2001) grouped risks into four categories: environmental, industrial, client, and project. Shen (2001) classified risks by nature of occurrence, such as financial risk, market risk, management risk, legal risk, and political or regulatory risk. In Chen et al. (2004), there are just a few risk factors that affect construction costs, which were divided into three categories: resource, management, and parent. A study conducted by Dikmen et al. (2007) employed influence diagrams to identify the factors that have an impact on project cost. Risk factors were further classified by Zeng et al. (2006) into four categories: human factor, site factor, material factor, and equipment factor.

In recent times, more sophisticated techniques of risk management have been developed. Table 8.1 shows the combination of several factors experienced in risk and that of the categories by which it is divided as stated by the authors. Rezakhani (2012) classified risk into three types: external, legal, and internal. It was determined that external risk could be further classified into two: unpredictable/uncontrollable and predictable/uncontrollable, while internal risk could also be further classified into non-technical/controllable and technical/controllable. In addition, the author proposed a risk breakdown structure based on a hierarchy of risks and identified the most significant risk factors. Goh et al. (2013) classified the numerous risk factors into five stages: planning stage, design stage, procurement stage, construction stage, and handing-over stage. The authors identified the usage of workshops with an integrated approach, brainstorming, checklists, probability impact matrices, subjective

Table 8.1 Techniques in risk management

Author	Techniques in RM propounded
Rezakhani (2012)	Risk in terms of categories: External risk Internal risk Legal risk
Goh et al. (2013)	Risk in terms of factors: Planning state Design stage Procurement stage Construction stage Handling-over stage

judgment, and risk register as ways of improving awareness of risk management practice. Lastly, the study recommended that the approach of using risk management workshops will be more beneficial for risk identification and analysis.

Past studies have made numerous attempts to illustrate the interdependences between project risks and the complexity of the surrounding environment using a variety of methodologies (Lazzerini & Mkrtchyan, 2011). In Hwang et al. (2013), the implementation of risk management in small-scale construction is often considered low, and this is due to a lack of time and budget, low-profit margin, and the fact that it is not cost-effective. The findings demonstrated a positive association between the applications of risk management practice toward improvement in construction quality, cost, and whole life cycle.

The History of Risk Management in Tanzania Construction Sector

The role of the construction sector cannot be overemphasized, as it influences socio-economic development in various ways (Ofori, 2007). This impact is particularly significant in developing nations because of the strong need for infrastructural development and housing. As a result, the construction sector actively promotes economic and social development, leading to wealth, social equality, and improved standard of living (Gomes & Silva, 2005). An example of this is seen in the study by Lema (2008) where it was stated that the construction industry in Tanzania contributes approximately 7% of the country's GDP and further corroboration in the study by Kikwasi (2013).

In order to optimize the performance of the construction sector, it is necessary to analyze the causal relationship between the implementation of certain risk management measures and the success of a project (Tabish & Jha, 2011). From this, it can be established that enhanced knowledge of risk management practice and its subsequent application could contribute to overall project performance optimization in the construction sector.

Research findings indicate that most construction firms in Sub-Saharan Africa do not incorporate risk management strategies and methodologies into their project management procedures, which frequently results in project cost and time overruns (Kululanga & Kuotcha, 2010; Kikwasi, 2013). As a result, it is necessary to investigate the obstacles that are preventing the execution of risk assessment and management practices in developing nations using Tanzania as an example.

The following is an account of some of the issues that the Tanzania construction sector has experienced. Here, we explain explicitly the challenges of adopting and implementing RAMP based on existing literature and then present a concise summary of discussions and information on the knowledge gap that exists. The methodological technique used in the research, as well as a discussion of the study's findings and implications, is presented briefly. Some solutions to the barriers to RAMP implementation have been advocated, and some of these have been proposed.

Prevailing Challenges of the Tanzanian Construction Industry

Tanzania aspires to have one of the top construction industries in the world but continues to struggle with poor project execution (Ofori, 2012), a shortage of trained experts, and other challenges (Debrah & Ofori, 2005). Tanzania relies heavily on international institutions to train indigenous experts, contractors, and consultants for large-scale construction projects on her territory (Debrah & Ofori, 2005; Egmond, 2012).

These prevailing challenges have negatively impacted the general adoption and implementation of project management concepts and principles among the many parties involved in the project (Lema, 2008). Several studies have been conducted in Tanzania on project management-related topics such as procurement and project performance, but none have explicitly addressed risk management challenges that arise during the project development process.

In a bid to bridge this gap, Kikwasi (2013) conducted a study to determine the causes, impacts, and disruptions of construction projects in Tanzania. According to the findings of this study, there are still various factors contributing to risk management delays and disruptions. These effects put construction projects at serious risk of failure and negatively impact their performance. Poor project management was identified as one of the causes of project failure. It is well acknowledged that international contractors are involved in the construction industry operational environment in the mentioned country. These international contractors often have an edge over local contractors due to their superior skills, training, and competences, as well as their human resources development (HRD) practices. For instance, the Constructors Registrations Board (CRB) (2010) estimates that there are about 4470 registered contractors in Tanzania, with 134 (3%) from abroad and 4336 (97%) from within the country. Despite the small number of foreign contractors in the construction industry, they control approximately 96% of the market share. Local contractors should be encouraged and trained to meet the international standard and, from there would be able to deliver quality projects.

Impediments to the Adoption of Risk Management Practice in Tanzania

The adoption and implementation of risk management practices have been the subject of a number of research studies. Even though each country's construction faces similar and different impediments to the practice, it can be concluded that the impediments are mostly similar, especially in most developing countries.

In the research conducted by Frimpong et al. (2003), which sought to uncover the causes of cost and time overrun in the Ghanaian construction industry, one of the factors identified was limited knowledge of risk management practices.

In a further study, Akintoye and MacLeod (1997) discovered that contractors' lack of technical knowledge of risk analysis and management was one of the reasons for not employing risk management techniques. However, Wood and Ellis (2003) discovered that one of the elements limiting the implementation of risk management practices was a relative lack of training and skill development.

Chileshe and Kikwasi (2014) identified some of the impediments to risk management implementation in the Tanzanian construction industry. These include but are not limited to:

- Lack of understanding of the risk management process.
- Inexperience in process execution.
- Lack of information.
- Lack of proper coordination among parties involved.
- Availability of specialist risk management consultants.
- Cost implications.
- Time constraints.

Framework/Model for Risk Management in Tanzania

Risk management system is a process that comprises risk analysis and risk response. Risk analysis is subdivided into risk identification and assessment.

Risk Identification

This is the first step in risk mitigation, it involves the use of techniques and tools to brainstorm and break down the inherent risk (Maytorena et al., 2005). In the quest to identify risks associated with a particular project, it is, however, important to identify the different sources of these risks and also their effects on the project. This would enable one to set priorities right and proffer suitable solutions (Flanagan & Norman, 1993).

Risk Assessment

This is a process that entails employing proper tools to ascertain the likelihood of an event occurring and the expected impact. Risk assessment will enable the risk manager to categorize the unforeseen events within a range of highest probable to the least probable (Grey, 1995).

Risk Response

This is the process of attempting to offset the impacts of a risk. Risk response is typically categorized into four levels: risk retention, risk reduction, risk transfer, and risk avoidance (Flanagan & Norman, 1993).

Conclusion

The Tanzanian construction industry has experienced improvement over the years. Through the implementation of favorable policies and frameworks to develop the country's construction sector, project planning and execution have increased tremendously as construction professionals are more knowledgeable than before. However, the country's construction sector is still largely dominated by foreign contractors. This has been attributed to their level of skills, training, education, sex, etc., in handling projects irrespective of the size, nature, and type. The implementation of risk management practices through an increased number of local contractors will help the country to develop better and thus increase overall development.

References

Akintoye, A. S., & MacLeod, M. J. (1997). Risk analysis and management in construction. *International Journal of Project Management, 15*(1), 31–38.

Bannerman, P. L. (2008). Risk and risk management in software projects: A reassessment. *Journal of Systems and Software, 81*(12), 2118–2133.

Birnie, J., & Yates, A. (1991). Cost predictions using decision/risk analysis methodologies. *Construction Management and Economics, 9*(2), 171–186.

Chapman, R. J. (2001). The controlling influences on effective risk identification and assessment for construction design management. *International Journal of Project Management, 19*(3), 147–160.

Chen, H., Hao G., Poon, S. W., & Ng, F. F. (2004). Cost risk management in west rail project of Hong Kong. In *AACE International Transactions*. INT.09.1-INT.09.5, China.

Chia, E. S. (2006). *Risk assessment framework for project management.* Retrieved from https://www.ieeexplore.ieee.org

Chileshe, N., & Kikwasi, G. J. (2014). Critical success factors for implementation of risk assessment and management practices within the Tanzanian construction industry. *Engineering, Construction and Architectural Management, 21*(3), 291–319.

Constructors Registrations Board (CRB). (2010). *Constructors registrations board consultative meeting, capacity and competence of contractors' – A key to sustainable infrastructure development.* Colour Print (T) Limited.

Debrah, Y. A., & Ofori, G. (2005). Emerging managerial competencies of professionals in the Tanzanian construction industry. *The International Journal of Human Resource Management, 16*(8), 1399–1414.

Dikmen, I., Birgonul, M., & Han, S. (2007). Using fuzzy risk assessment to rate cost overrun risk in international construction projects. *International Journal of Project Management, 25*(5), 494–505.

Douglas, H. (2009). *The failure of risk management: Why it's broken and how to fix it*. Wiley.

Egmond, E. V. (2012). *Case studies of construction technology development and innovation in developing countries: Contemporary issues in construction in developing countries*. Spoon Press, Taylor & Francis.

Flanagan, R., & Norman, G. (1993). *Risk management and construction*. Blackwell Scientific Publications.

Frimpong, Y., Oluwoye, J., & Crawford, L. (2003). Causes of delay and cost overruns in construction of groundwater projects in a developing countries; Ghana as a case study. *International Journal of Project Management, 21*(5), 321–326.

Goh, C. S., Abdul-Rahman, H., & Samad, Z. A. (2013). Applying risk management workshop for a public construction project: Case study. *Journal of Construction Engineering and Management., 139*(5), 72–580.

Gomes, V., & Gomes da Silva, M. (2005). Exploring sustainable construction: implications from Latin America. *Building Research & Information, 33*(5), 428–440. https://doi.org/10.1080/09613210500218891

Grey, S. (1995). *Risk assessment for project management*. Wiley.

Hwang, B. G., Zhao, X., & Toh, L. P. (2013). Risk management in small construction projects in Singapore: Status, barriers and impact. *International Journal of Project Management, 32*(1), 116–124.

Karimi, A., Mousavi, N., Mousavi, S. F., & Hosseini, S. (2011). Risk assessment model selection in construction industry. *Expert Systems with Application, 38*(8), 9105–9111.

Kikwasi, G. J. (2011). An assessment of risk management practices by consultants in Tanzania. In *Proceedings of the 6th built environment*, Johannesburg.

Kikwasi, G. J. (2013). Causes and effects of delays and disruptions in construction projects in Tanzania. *Australian Journal of Construction Economics and Buildings, 1*(1), 52–59.

Kululanga, G. K., & Kuotcha, W. S. (2010). Measuring project risk management process for construction contractors with statement indicators linked to numerical scores. *Engineering, Construction and Architectural Management, 17*(4), 336–351.

Laryea, S., & Hughes, W. (2008). How contractors price risk in bids: Theory and practice. *Construction Management and Economics, 26*(9), 911–924.

Latham, M. (1994). *Constructing the team*. Retrieved from https://www.scirp.org

Lazzerini, B., & Mkrtchyan, L. (2011). Analyzing risk impact factors using extended fuzzy cognitive maps. *Institute of Electrical and Electronics Engineers (IEEE) Systems Journal, 5*(2), 288–297.

Lema, N. M. (2008). *Project management practice in engineering performance and future challenges*. University Press.

Mark, W. Cohen, P. E., & Glen, R. P. (2004). *Project risk identification and management*. Retrieved from https://www.proquest.com/openview

Maytorena, E., Winch, G. M., & Kiely, T. (2005). Construction risk identification. In *Proceedings of the 11th joint CIB international symposium – Combining forces advancing facilities management and construction through innovation*. 13–16 June, Helsinki.

Ofori, G. (2007). Construction in developing countries. *Construction Management and Economics, 25*(1), 1–6.

Ofori, G. (2012). *The construction industries in developing countries; strategic review of the book*. Retrieved from https://www.taylorfrancis.com

Project Management Institute (PMI). (2008). *A guide to the Project Management Body of Knowledge (PMBOK guide)*. Project Management Institute Incorporation.

Rezakhani, P. (2012). Fuzzy MCDM model for risk factor selection in construction projects. *Engineering Journal, 16*(5), 79–94.

Shen, L. Y., Wu, G. W., & Ng, C. S. (2001). Risk assessment for construction joint ventures in China. *Journal of Construction Engineering and Management., 27*(1), 76–81.

Tabish, S. Z., & Jha, K. N. (2011). Identification and evaluation of success factors for public construction projects. *Construction Management and Economics, 29*(8), 809–823.

Tuysuz, F., & Kahreman, C. (2006). Project risk evaluation using a fuzzy analytic hierarchy process: An application to information technology projects. *International Journal of Intelligent Systems, 21*(6), 559–584.

Wood, G. D., & Ellis, R. C. T. (2003). Risk management practices of leading UK cost consultants. *Engineering, Construction and Architectural Management, 10*(4), 254–262. https://doi.org/10.1108/09699980310489960

Zeng, J., An, M., & Smith, N. J. (2006). Application of a fuzzy based decision making methodology to construction project risk assessment. *International Journey of Project Management, 21*(6), 589–600.

Part III
Risk Management in Selected Developed Countries

Chapter 9
Risk Management Practices in Australia

Abstract Risks that have not been identified and handled pose an undeniable threat to the project's objectives, potentially resulting in major cost and time overruns. Risk management is a proactive decision that entails several steps, including risk identification, analysis, risk response and communication, monitoring, review, and learning. This chapter focused on risk management practice in Australia by providing guidelines and principles for establishing an effective risk management framework. The chapter also provided the finest risk management framework and process for Australian construction projects, which was adapted from AS/NZS ISO 31000:2009. The concluding sections of the chapter gave the benefits and other related topics pertaining to the Australian construction industry.

Keywords Construction risk · Project risk · Risk analysis · Risk communication · Risk identification · Risk response

Introduction

The construction industry in Australia has developed to become one of the most significant in the country's economy. In 2015, the industry generated more than AUD 120 billion, accounting for 8% of Australian industry's gross value added. The construction industry has proven to be one of the most active sectors of the economy due to its overall impact on the country's economic development and its populace (Tabish & Jha, 2011).

The complicated nature of construction projects exposes them to constant risks. Risk is referred to as the probability of any event happening that will have an impact on the project aim and objectives. According to the Australian Standard Risk management (2004), risk management is the culture processes and structures aimed at maximizing possible possibilities while minimizing negative consequences. Risk management involves certain processes such as risk identification, analysis, response, communication and monitoring, review, and learning. Monetti et al. (2006) opined that risk management is essential for achieving organization or

A. E. Oke et al., *Risk Management Practices in Construction*,
https://doi.org/10.1007/978-3-031-35557-8_9

project objectives since it is not only aimed at avoiding negative outcomes. Also, it can act as a guide to maximize favorable outcomes within the concept of available resources and practices.

Atkinson (2001) reported that construction projects are all prototypes to some extent, facilitated by changes, distinctiveness, and similarities. Zhi (1995) further corroborates that each project is distinct and thus invariably entails complex and diverse risks. The nature of risks influences the duration of a construction project (Sigmund & Radujkovic, 2014). In Marsh Australia (2007), risks are exacerbated by certain factors, which include legislative changes, the influence of relevant authorities, the use of non-standard building contracts, and unpredictable site conditions.

Construction projects involve various construction methods and several team participants with varied skills and backgrounds. The lack of adequate risk management practices among these team members can result in inefficient use of labor and resources, which subsequently affects project execution. Loosemore et al. (2006) highlighted that the negative effects of a lack of risk management process could result in project outcome uncertainty, financial crisis, liabilities, and poor decision-making, among other effects.

To identify goals and processes and improve risk management, organizations must have a thorough understanding of the current and functional risk management approach available within the context of the firm (Risk Management Research and Development Program Collaboration RMRDPC, 2002). In Lyons and Skitmore (2004), risk management is more often used at the planning/design stage of the project life cycle and also throughout the construction stage than in the post-construction stage. The authors identified risk identification and risk assessment as the most frequently used risk management processes, followed by risk response and risk documentation.

Brainstorming is considered to be the most commonly used tool for risk identification. For risk assessment, a qualitative approach is the most commonly used, while risk reduction is the most commonly used risk response method, with contingencies and contractual transfer preferred over insurance; project teams are the most frequently used group for risk analysis, outnumbering in-house specialists and consultants.

To attain a high degree of performance within an organization or on a construction project, the issue of risk must be addressed, and the most effective risk management practice is identified.

History of Risk Management in the Australian Construction Industry

In the study by Al-Bahar (1988), it was stated that many contractors developed a set of guidelines for analyzing and assessing risks. This was due to numerous construction projects failing to meet schedule, cost, and quality objectives. It was proposed that during the estimation stage, a systematic method for managing project risks and uncertainties in cost and time may be employed to reduce their effects.

Unmanaged or uncontrolled risks are one of the leading causes of project failure; hence, risk management is an important aspect of project management. In the early stages of risk management development, even though there were several published works in the area of risk management, there was little information available on how the practice could be applied. Between 1987 and 1997, various surveys on risk management were conducted in a variety of Western nations in the United States, the United Kingdom, Saudi Arabia, Australia, Canada, and Israel. Even though the results improved with time, its objectives were challenged heavily at the initial stage.

The conceptual stage is critical for a new construction project since decisions taken during this phase have a substantial influence on the final cost. It is also the stage in which the greatest amount of ambiguity regarding the future is met. In a situation like this, risk management may play an essential role in regulating the level of risks and limiting their impacts. However, excluding projects that involved high risk, such as we have in the petrochemical, oil exploration, and aerospace industry mentioned by Flanagan and Norman (1993) and Toakley (1998), its adoption by the Australian construction industry was rather slow.

Ward et al. (1991) hypothesized in their study the necessary preconditions for successful risk management implementation in construction, which are the design of a suitable organizational structure tailored toward supporting the basic performance standards of risk management, as well as a new organizational culture to drive its implementation.

In 1993, the New South Wales (NSW) Government of Australia took a significant step by mandating the use of risk management in the planning of new projects and large capital asset operations costing more than $5 million (Flanagan & Norman, 1993). This document states that risk management should start at the strategic planning phase and continue through the project's life cycle.

Australian Risk Management Standard

The Australian and New Zealand risk management standard AS/NZS 4360: 2004 was replaced by AS/NZS ISO 31000: 2009 in November 2009. AS/NZS ISO 31000:2009 provides foundational member agencies with general guidelines and principles to consider when drafting risk management frameworks and procedures. Risk management is defined by the Australian New Zealand Risk Management Standard (AS/NZS ISO 31000:2009) as an organized effort to guide and control an organization with reference to risk. The Australia Council for the Arts ("the Australia Council") faces risk in all elements of its operations and at all stages of its life cycle. The planning and execution of the council projects provide both opportunities and threats and therefore need to be managed properly. ISO31000:2009 has recently been amended, giving rise to ISO31000:2018 described in Sousa et al. (2012). The primary improvements here include a review of risk management principles, highlighting the organization's leadership, putting more emphasis on the iterative nature of risk management, and tailoring the content to accommodate diverse demands and situations.

The risk management obligations of the Australia Council are outlined in the Public Governance, Performance and Accountability Act (PGPAA) 2013. This requires the Australia Council to design and maintain the following:

1. An effective risk monitoring and management system for the organization.
2. An adequate internal control system for the organization.

The following are the risk management principles as stated by the Australian Council.

- Creates and protects value.
- Be an important element of organizational processes.
- Participates in decision-making.
- Be methodical, structured, and timely.
- Be transparent and inclusive.

To guarantee alignment with risk appetite and organizational objectives, risk management is included in the Australia Council's strategic and operational planning procedures at all levels, including all new activities, initiatives, and projects before they are implemented.

Risk is captured and reported as follows, according to the Australia Council.

1. *Risk Management Objective Profile*

This is a hierarchical overview of the Australia Council's main risk categories, objectives, related procedures, and residual risk areas uncovered through strategic planning processes, emergent risks, and a comprehensive risk profile review. The new risk management framework policy calls for a bi-annual evaluation of the risk profile by members of the executive team in collaboration with the audit and finance committee to ensure that existing sources of risks and their associated action plans are clearly acknowledged and accounted for.

2. *Detailed Risk Register*

A detailed register shows the major area of inherent and control risks attached to the operation procedures that are relevant to the attainment of the Australia Council's risk management objectives. It includes comprehensive risk evaluation rating, control effectiveness ratings, and action plans. In conjunction with its management, the detailed risk register is examined and offers a framework for assessing residual risk and specific risk control areas where additional review, risk management action plans, and internal auditing are required.

3. *Risk Assessments for Specific Projects*

Risk assessment and registers tailored to specific projects ensure risks connected with significant strategic projects or regulatory obligations should be properly managed.

Process of Risk Management Assessment

To identify and manage risks, the best risk management practice/framework in AS/ NZS ISO 31000:2009 Risk Management Process will be adopted.

The essential components of the risk management process, as illustrated in Fig. 9.1 and Table 9.1 are described below.

Establishing the Context—The Australian Council analyses both external and internal factors when evaluating and managing risks connected with the fulfillment of organizational objectives.

Risk Assessment—It includes full risk identification, analysis, and evaluation process.

Risk Identification—It is the process of identifying risk sources, areas of influence, events, causes, and potential outcomes in order to compile a comprehensive list of risks based on occurrences that could produce, enhance, prevent, degrade, speed up, or delay the fulfillment of set out goals.

Risk Analysis—It is the process of determining a risk rating by taking into account a variety of causes, sources of risk, consequences, and likelihood. The Australian Council can utilize the rating to determine future management.

Risk Evaluation—It is the process that involves using a consistent overall ranking and rating system to rate and prioritize the level of risk detected during risk analysis.

Communicate and Consult—Effective communication, consultation, and risk management education are essential to accomplish a successful integration of risk procedures within the organization.

Risk Treatment—It includes deciding on one or more alternatives for mitigating risks, taking into account other resources available.

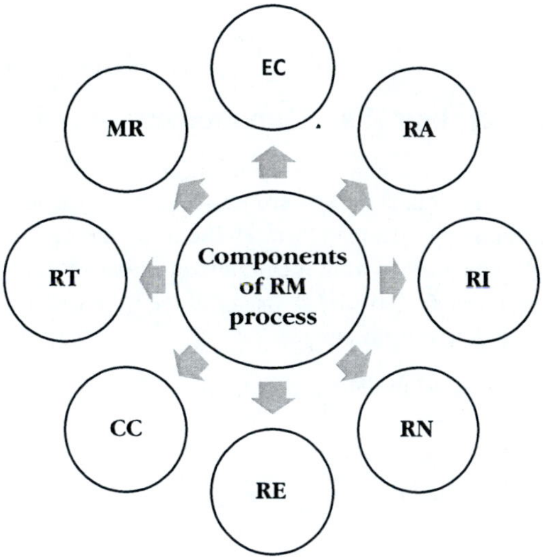

Fig. 9.1 Components of risk management process in construction

Table 9.1 Interpretation of the components of risk management (RM) process in construction

Code	Meaning
EC	Establishing the context
RA	Risk assessment
RI	Risk identification
RN	Risk analysis
RE	Risk evaluation
CC	Communicate and consult
RT	Risk treatment
MR	Monitoring and review

Monitoring and Review—The efficacy and appropriateness of the Australian Council risk management profiles, which include risk treatment plans, risk assessments, and the detection of developing risks, must be monitored and examined on a regular basis.

The probability and consequence levels are used to determine the level of inherent risk. The control effectiveness rating is then determined by evaluating the mitigating procedures and controls associated with the inherent risks. The residual risk rating and treatment plan are calculated using the combined inherent risk rating and control effectiveness. The residual risk is the degree of risk that remains within the council after considering all existing mitigating practices/controls. The residual risk indicates the level of management attention that is required, as well as when treatment plans are needed to ensure the management of such risk.

Summarily, the executive team of the Australian Council is saddled with the responsibility of risk management and how risk controls are identified, implemented, and monitored. The Executive team will advise the Executive Director of Corporate Resources and the designated Risk Manager on issues of strategic and operational relevance relating to risk identification and management.

Australian Risk Management Standard Framework

The risk management framework is a supporting structure that ensures risks are effectively managed through the implementation of the risk management process at various levels within the organization-specific context. According to the most recent ISO 31000: 2018 by Sousa et al. (2012), risk management framework components include the following:

- Leadership and commitment.
- Design.
- Implementation.
- Evaluation.
- Improvement.

- Integration.

Leadership and Commitment The leadership and regulatory bodies of the organization should ensure that risk management is implemented at all levels and should demonstrate leadership and commitment by the following:

- Ensuring that the risk management policy is defined and implemented.
- Ensuring that the organization's culture is consistent with the risk management policy being implemented.
- Making certain that the necessary resources are devoted to risk management.
- Ascribing responsibility and maintaining compliance at all levels.

Integration Understanding organizational structures and context is required for integrating risk management. Structures fluctuate depending on the organization's purpose, aims, and complexity. Risk is managed throughout the structure of the organization.

Design Certain procedures are involved in the process of designing the framework for risk management.

1. *Understanding the Organization and Its Environment:* Before Developing or Executing a Risk Management Framework, it Is Necessary to Understand the External and Internal Contexts of the Organization

   Understanding the organization's external context entails understanding:

   - Its social and cultural, political, financial, legal, regulatory, and economic environments, whether international, national, regional, or local.
   - The primary factors and trends that have substantial impacts on the objectives of the organization.
   - Relationships, views, attitudes, requirements, and expectations of external stakeholders.
   - Contractual connections and commitments.
   - Network and dependency complexity.

   While understanding the organizations, the internal context includes understanding its:

   - Vision, mission, and values.
   - Organizational structure with leadership at all levels and their attributed roles and responsibilities.
   - Strategy, objectives, and policies.
   - Culture.
   - Standards, guidelines, and models.
   - Capabilities, as defined by resources and knowledge (e.g., capital, time, people, intellectual property, processes, systems, and technologies).
   - Data, information systems, and information flows.
   - Relationships with internal stakeholders, taking their perspectives and values into account.

- Contractual obligations and responsibilities.
- Interconnections and interdependence.

2. *Clearly Expressing Risk Management Commitment*: Where applicable, top management and regulatory bodies should demonstrate and define their unwavering commitment to risk management through policy, statement, or other forms that clearly convey the objectives and commitment of the organization to risk management.
3. *Delegating Organizational Roles, Authority, Responsibilities, and Accountability*: Where applicable, the leadership and regulatory bodies of the organization should ensure that authorities, responsibilities, and accountability for relevant risk management roles are allocated and communicated at all levels of the organization. Also, they should underline the importance of risk management as a primary duty and also determine who is saddled with the responsibility and authority to manage risk, also known as risk owners.
4. *Allocating Resources*: The regulatory bodies and the leadership of the organization should oversee the deployment of adequate risk management resources, which include the following:
   - People, skills, expertise, and competence.
   - The risk management processes, methods, and tools of the organization.
   - Defined processes and procedures.
   - Information and knowledge management systems.
   - Requirements for professional development and training.
5. *Establishing Communication and Consultation Mechanism Within the Organization*: To support the framework and effective implementation of the risk management process, the organization should guarantee that there is a regular mechanism for communication and consultation within the organization. Communication entails the exchange of information by using all the means available. Consultation entails formally providing comments to seek recommendations for needed improvements. Communication and consultation should be timely, with important information being collected, aggregated, synthesized, and shared as needed, as well as feedback being provided and improvements being addressed.

Implementation The designed risk management process must be implemented at all relevant levels of the organization. This can be accomplished by the following:

- Developing an efficient approach that involves time and resources.
- Determining where, when, and how various sorts of decisions are made across the organization.
- Modifying the applicable decision-making procedures as needed.
- Ensuring that risk management arrangements of the organization are well understood and practiced.

Evaluation To evaluate the effectiveness of an already implemented risk management framework, the leadership or regulatory bodies of the organization must measure risk management framework performance against its objective, implementation plans, indicators, and expected behavior on a regular basis. Also, they must determine whether the framework is still appropriate to help the organization achieve its objectives.

Improvement To address external and internal changes, the organization must constantly assess and adjust the risk management framework. There is also a need to continuously improve the framework suitability and effectiveness and the way the process is integrated.

Benefits of Risk Management

Effective risk management involves quantifying uncertainties that might develop during project execution, identifying the source of the uncertainties, and determining the potential consequences of such uncertainties. The implementation of risk management early in a project clarifies project objectives and provides significant cost savings.

Mills (2001) summarized some of the major benefits of systematic risk management on a project as it:

- Examines the assumptions that have the most impact on the project's success.
- Focuses on activities that will best control risks.
- Determines the cost–benefit of such activities.

Challenges of Risk Management

In most parts of the world, including Australia, the implementation of enterprise risk management within an organization, which ensures that risks are managed holistically, is strongly affected by a variety of factors. These factors are as follows:

1. *Lack of Knowledge or Understanding of Risk Management*: To be able to understand, use, and interpret the risk management process, there is a need for staff and managers to have some basic knowledge of statistics and other quantitative methods of analysis. Therefore, managers need to undergo necessary training, workshop, and education on risk management techniques to be aware of some of the benefits this process can offer.
2. *Risk Culture Resistance*: According to Dornberger et al. (2014), organizations must foster a risk culture in which risk and its management become institutionalized. This makes employees to be aware of the organization goals, engage in risk management, understand the significance of their actions, and take responsibility for the risks that are peculiar to their area of expertise. This

can be a stressful process that requires people in an organization to change their usual way of doing things.

3. *Lack of Support from the Top and Middle Managers*: In a situation where the executive managers do not see the need for the organization to have a defined approach to risk management, this creates a challenge in the implementation of this particular process. Also, this sort of attitude can be rubbed off on staff, supervisors, and managers who are to ensure that the risk management process is implemented. Therefore, to achieve effective implementation, there is need for the top and middle managers to be in support of the risk management efforts presented by the organization.
4. *Nature of the Environment*: To implement risk management, it is necessary to first understand the nature of the environment and the world in which we live, as it can be uncertain, complex, and volatile all at once. This is due to changes in natural phenomena as well as changes in the political, economic, and social environment.

Conclusion

Construction is a risky business; thus, a working risk management policy is essential both before and throughout the construction phases. The chapter analyzed numerous articles of literature on the subject and assessed the history of risk management in Australia, and the government provided guidelines to enable effective risk management implementation at all levels of the organization. The chapter adopted the most recent risk management framework defined in ISO 31000: 2018 to further assess the practice in the mentioned country.

References

Al-Bahar, J. F. (1988). *Risk management in construction projects: A systematic analytical approach for contractors.* Ph.D. Dissertation, Department of Civil Engineering, University of California.

Atkinson, D. (2001). *Risk management in construction projects*. Retrieved from http://www.atkinson-law.com/cases/casesArticles/Articles/Risk_Management.htm

Dornberger, K., Oberlehner, S., & Zadrazil, N. (2014). Challenges in implementing enterprise risk management. *Journal of Finance and Risk Perspectives, 3*(3), 1–14.

Flanagan, R., & Norman, G. (1993). *Risk management and construction*. Blackwell Scientific Publications.

Loosemore, M., Raftery, J., Reilly, C., & Higgon, D. (2006). *Risk management in projects* (2nd ed.). Taylor and Francis.

Lyons, T., & Skitmore, M. (2004). Project risk management in the queensland engineering construction industry: A survey. *International Journal of Project Management, 22*(1), 51–61.

Marsh Australia. (2007). *Construction*. Retrieved from http://www.marsh.com.au/Construction.htm

Mills, A. (2001). A systematic approach to risk management in construction. *Structural Survey, 19*(5), 245–252.

Monetti, E., Rosa da Silva, S. A., & Rocha, R. M. (2006). The practice of project risk management in government projects: A case study in Sao Paulo City. In *Proceedings of CIB W107 construction in developing countries international symposium,* 18–20 January, Santiago.

Public Governance Performance and Accountability (PGPA). (2013). *Public governance performance and accountability act 2013.* Retrieved from https://www.legislation.gov.au

Risk Management Research and Development Program Collaboration. (2002). *Risk management maturity level development.* Retrieved from https://www.pmiswitzerland.ch/fall05/riskmm.pdf

Sigmund, Z., & Radujkovic, M. (2014). Risk breakdown structure for construction projects on existing buildings. *Procedia – Social and Behavioral Sciences, 119*(3), 894–901.

Sousa, V., Almeida, N. M., & Dias, L. A. (2012). Risk management framework for the construction industry according to the ISO 31000:2009 standard. *Journal of Risk Analysis and Crisis Response, 2*(4), 261–275.

Tabish, S. Z., & Jha, K. N. (2011). Identification and evaluation of success factors for public construction projects. *Construction Management and Economics, 29*(8), 809–823.

Toakley, A. R. (1998). *Risk, uncertainty and subjectivity in the building procurement process – A critical review.* School of Building, University of New South Wales.

Ward, S. C., Chapman, C. B., & Curtis, B. (1991). On the allocation of risk in construction projects. *International Journal of Project Management, 9*(3), 140–147.

Zhi, H. (1995). Risk management for overseas construction projects. *International Project of Project Management, 13*(4), 231–237.

Chapter 10
Risk Management Practices in Canada

Abstract Risk plays a significant part in getting a successful outcome in a construction project. It is an element that must be appropriately considered during the project planning process. Risk management, the study of risk, begins with the proactive identification of prospective risks and progresses to the continuing management of those risks deemed appropriate. Hazards and risks are terms that are used interchangeably; this chapter provided insight into events that might be classified as hazards or risks, depending on the circumstances. Risk has gained traction in almost every province in Canada, but there is still no defined strategic approach or framework for successful risk management. The chapter further examined the various approaches to effective risk management by providing a holistic risk management framework that will enable a more citizen-centered approach to improving public interest decision-making and a project outcome. Also, some of the benefits of risk management implementation in Canada were explained, which were sourced from the Treasury Board of Canada publications.

Keywords Construction risk · Project risk · Project success · Risk evaluation · Risk identification · Risk management

Introduction

Risk management has been identified as a core element for public administration and project management by assisting in better decision-making, ensuring better resource allocation, and ultimately producing a better result for the people or organization. Risk is defined by the Society for Risk Analysis (2022) as the potential for realizing undesirable, unfavorable effects on humans, health, property, or the environment. To manage risks, organizations and groups of all types and sizes coordinate activities that provide direction and control as far as risk is concerned. According to the International Organization for Standardization (ISO) (2022), risk assessment (RA) is a key element of risk management, which encompasses risk identification, risk analysis, and risk evaluation. Risk assessment allows organizations

A. E. Oke et al., *Risk Management Practices in Construction*,
https://doi.org/10.1007/978-3-031-35557-8_10

to gain an enhanced understanding of risk, its causes, consequences, and probabilities. International Electro-Technical Commission (IEC) (2018) stated that the acquired knowledge by the organizations is used to "offer evidence-based information and analysis to make smart choices on how to handle specific risks and how to pick among solutions."

Canadian Construction Industry

The construction industry in Canada is a significant contributor to the country's economy, accounting for 6% of the country's gross domestic product (GDP) (Industry Canada, 2012). In 2013, the industry provided 1.2 million jobs to the economy, accounting for 7.1% of the total labor force in the country (Human Resources and Skills Development Canada, 2013). The construction industry's annual expenditure is expected to increase from $7.2 trillion in 2011 to $12 trillion by 2020, accounting for 13.2% of global GDP (Betts et al., 2011). According to Statistics Canada, between 2000 and 2010, the construction sector experienced a 42.7% increase in GDP, compared to a 20.2% increase for all sectors.

Canada's immigration policy allows for openness and minimizes discrimination, allowing any organization to strive and ultimately promote innovation, and this is one of the factors that distinguish the construction industry. Despite the industry's massive growth in recent times, there is still a massive gap between the number of jobs available and the available number of skilled laborers that need to be bridged due to the ever-growing population, which drives demand for residential and commercial building projects.

Risk Management in Canada

Risk management has been recognized as a critical management principle that has often been articulated in terms of its role in "taming chance" through the assessment and control of uncertainty. Traditionally, risk management in organizations has been conducted in discrete sectors, such as technological, legal, commercial, and environmental, with little coordination. Furthermore, major risk events need that the structures, strategies, and processes for managing organizational risk be consistent and both internal and external organizational systems be reviewed on a regular basis (Grant & Venzin, 2009; McGee, 2005).

Risk Analysis and Assessment

This is a methodical technique for determining the criticality of recognized risks in a project. In Weaver (2008), where reference was made on the Project Management Institute (PMI) framework, the risk management process includes risk planning, risk identification, qualitative and quantitative assessment, risk response planning, risk monitoring, controlling, and reviewing.

Effective risk management necessitates a risk analysis methodology that is founded on a scientific framework and facilitated by quantitative and qualitative approaches. Quantitative risk analysis approaches are based on numerical methods, while qualitative risk analysis approaches prioritize identified risks based on the rate of occurrence and impact on project goals. The quantitative and qualitative analyses can both be conducted at the same time and are not mutually exclusive.

Some of the approaches used in qualitative risk analysis include Brainstorming, Delphi, Final project reports—lessons learned, probability and impact matrix, AHP (analytic hierarchy process) method, and root cause analysis of risk. The following are the approaches to quantitative risk analysis that have been recommended, but not limited to the following:

- Sensitivity analysis approach.
- Program Evaluation Review Technique (PERT).
- Probability analysis approach.
- Monte Carlo.
- Decision tree (The Treasury Orange Book, 2004).

There is a notion that to evaluate if a certain approach is appropriate for assessing a given project risk, the benefits and limitations of using the approach in terms of implementation under conditions specific to those projects must be determined. Assessing the benefits and limitations of risk analysis approaches allows the project team to easily choose a valid approach, eventually resulting in a better risk management process on a project.

Though risk assessment (RA) emerged in its applications to engineering and health (Paté-Cornell, 1996), it is now used across various fields, including public safety and security. As a result, RA can have great implications for the safety and security of affected populations, and it is critical to be educated in its usage and application. There is no unique method or approach for RA since the project is diverse per country and nature, but its diversified applications necessitate different methods that are mostly appropriate for the scope of the assessment and the availability of resources in relation to a targeted project planning and execution.

Qualitative and Quantitative Risk Assessment

One of the most widely debated issues in risk assessment is the choice between using qualitative and quantitative methods. The sections below give a brief description of some of the benefits and limitations of these two approaches.

Qualitative Risk Assessment Methods

Qualitative risk assessment (QRA) methods assign verbal ratings such as "high," "medium," or "low" to various dimensions of risk, such as probability and consequence (Genik & Chouinard, 2012). They are frequently used because they simplify risk assessments, reduce the required inputs and judgments, and can be easily communicated to policymakers and stakeholders (Cox et al., 2005). However, the simplicity of qualitative methods generates a variety of concerns with respect to mathematical accuracy and precision. We have three rating methods in qualitative research, they are as follows.

(a) *Frequency and Probability Ratings*

For frequency and probability ratings, Hartley (2018) explains that this qualitative risk assessment method uses a few intervals to represent a wide and continuous range of frequencies. This can be problematic, especially since it entails counting risks that might occur without an accurate comparative risk evaluation. In this method, the number of times a variable is responded to and the rate at which it can be responded to describe the context of its rating.

(b) *Consequence Ratings*

The problems discussed above apply equally to consequence ratings as stated by Hartley (2018). Cox (2008), in the study "What's Wrong with Risk Matrices," explains that there is no standardized way to categorize the comparative severity of events with highly uncertain consequences. Individuals have varying degrees of risk aversion (avoidance of risk), and these attitudes can change the way that humans order the consequence ratings of several events.

(c) *Risk Matrices*

Cox (2008) also explores the challenges and errors that can occur while using risk matrices. A risk matrix is a commonly used tool for assessing and communicating risk. It is a table that assigns categories for frequency on its rows and categories for a consequence on its columns (or vice versa).

Using the formula risk = probability × consequence, risk matrices permit users to determine various levels of risks for each row–column pair.

In the risk matrix described above, consequence and probability have numerical values ranging from 0 to 1 (inclusive). The quantitative risk for any pair (consequence, probability) is their product, as defined by the formula as follows:

$$\text{Risk} = \text{Consequence} \times \text{Probability}$$

In general, the risk matrix correctly classifies the two risks if one risk is in the "high" risk cell and the other is in the "low" risk cell. This is because every high cell risk is qualitatively and quantitatively greater than any risk that falls in the "low" risk cell. However, suppose that probability and consequence values are negatively correlated (i.e., as the consequence value increases, the probability value decreases) and that the risk pairs are clustered along the line probability = 0.75 – consequence, then for all of the risks that fall along this line, the risk classifications provided by the risk matrix do not accurately represent the true quantitative risks. Along this line, the risks that fall in the "medium" risk cells have *smaller* quantitative risks than the risks that fall in the "low" risk cells.

Quantitative Risk Assessment Methods

To overcome the limitations that emerge in qualitative risk assessment, suggestions have been made over time in using quantitative methods. In Kaplan and Garrick (1981), quantitative risk assessment was conducted by answering the following three questions:

1. What can happen?
2. What are the consequences?
3. What is the likelihood?

The first question is answered by identifying a set of initiating events or disturbances. Next, the consequences can be estimated using logic diagrams such as event and fault trees to generate event sequences (scenarios). Lastly, the likelihood of these scenarios is evaluated using all available evidence (primarily historical data) and expert judgment and assigned numerical rankings according to the expected frequency of occurrence (Apostolakis, 2004). The use of quantitative risk assessment is widespread because it employs logical and transparent mathematics and provides precise numerical values for risk (Cox et al., 2005). In addition, quantitative risk assessment also can analyze a wide range of possibilities and quantify uncertainty (Apostolakis, 2004). However, the reliability and accuracy of quantitative risk assessment can be challenged due to the numerous limitations and errors in the method. Different rating in quantitative methods is listed as follows.

(a) *Likelihood Ratings*

Likelihood ratings can be a source of human error and uncertainty. In quantitative risk assessment, likelihood is determined by using precise quantitative interval scales, such as very likely: >90%; likely: >66%; about as likely as not: 33–66%; unlikely: <33%; and very unlikely: <10%. However, such interval scales are often misinterpreted and erroneously applied in practice (Budescu et al., 2009).

(b) *Consequence Ratings*

The consequences of risk events are rated by using a variety of "impact factors," such as fatalities, economic damage, and environmental impact. However, estimating the impacts of potential scenarios creates several challenges because some impact factors are very difficult to quantify and measure. For example, how does one quantify the damage to national reputation or environmental degradation? In addition, impact estimates must be aggregated into a common metric to determine an overall risk rating for each scenario. However, the "impact factors" are very diverse and thus use different measurement units (such as the number of fatalities or the monetary loss in a particular currency). Consequently, the conversion of impact estimates into a common metric requires the use of subjective judgments. For example, risk assessors are often faced with the question, "what is the monetary value of a life"? (Hartley, 2018).

(c) *Scoring Methods and Ordinal Scales*

The consequence and likelihood ratings discussed above are frequently conducted using scoring methods with ordinal scales. However, Hubbard and Evans (2010) examined some of the limitations of this approach. On an ordinal scale, items are assigned numbers so that the order of the numbers represents the position or order of these items in relation to another. For example, on an ordinal likelihood scale of 0 to 5 (0 indicates no likelihood and 5 indicates very high likelihood), an item with a likelihood rating of 4 has a greater likelihood than an item with a rating of 2. Risk factors are rated using ordinal scales, which are then aggregated using additive weighting or multiplication so as to provide a measure of the overall risk.

Generally, the additive method is used to assess the overall risks of a project, policy, or investment. Using the additive method, each of these risk factors would be weighted according to their importance and then added to arrive at an overall risk score. In contrast, the multiplicative method is used to assess the risk of individual events. It is common to use scores for likelihood and consequence and multiply these values to give a final risk rating. This multiplicative method is in line with the generally accepted concept of risk: *risk = likelihood x consequence*. After providing an overview of the existing research on this method, the authors claimed that "scoring methods are not useful tools for risk assessment" (Hubbard & Evans, 2010).

Given the problems and limitations discussed above, Hubbard and Evans (2010) suggested that risk assessment should fulfill the three basic criteria:

- Risk assessment should use quantitative expressions of probability and the magnitude of losses. For example, instead of stating that the likelihood or impact is "medium" or "high," one should state, "10% chance of a loss of inventory is worth $2 million."

- Risk assessment should explicitly model the relationships and correlations between different system components. Instead of adding or multiplying risk factors, risk should be represented using descriptive functions.
- Risk assessment should apply methods to correct the biases in human judgment and decision-making.

Overview of Hazard and Risk

Risk and hazard are two words that are frequently used interchangeably. These two terms are of different meaning and do not really explains the same scenario. Wilson and McCutcheon (2003) in their work functionally described risk and hazard as follows.

Hazard This is a potential cause of danger or harm to person(s), environment, assets, or production during the usage of a machine, equipment, operation, material, or any other physical element being used in the working environment.

Risk This simply refers to the possible harm, loss, or environmental damage caused by a hazard. The likelihood of an unexpected incident and the severity of the resulting consequences define the relevance of risk. There are a few things worth mentioning about these definitions:

- *Risk occurs as a result of hazards.* Successful risk management requires thorough hazard identification; in other words, it is impossible to effectively manage risk posed by hazard without first identifying the source of the hazard.
- *There are four groupings of areas that might be affected by harm or damage: people, environment, capital (such as equipment, property,* etc.*), and production process*. The identification of these distinct groups leads to a more comprehensive risk management strategy that considers all possible losses. A wide range of engineering practices and possible hazards are included in comprehensive risk management.
- *Likelihood and severity are the two dimensions to risk.* Instead of *likelihood,* the term *probability* or *frequency* is sometimes used. Although the meanings of these terms differ slightly, particularly between frequency and probability, the difference is, however, not significant.

Risk management is described as the act of assessing exposure to loss and taking necessary measures to eliminate the risk or minimize it to a tolerable level. In other words, risk management refers to the whole process of identifying risk, assessing risk, and making choices to guarantee that proper risk controls are in place and being used.

Risk management starts with the proactive identification of potential risks, which lead to ongoing management of risks considered appropriate. This is the cycle of risk analysis, which contributes to risk assessment, which in turn contributes to risk management (Bird & Germain, 1996). Engineers basically analyze risk (in terms of likelihood and severity) in order to assess risk (in terms of acceptability) so that they can eventually control the risk. As previously explained, beginning this cycle without first accurately identifying the threats that must be addressed is simply not realistic.

Types and Sources of Risk Involved in Construction

There are different types/sources of risk involved in the construction industry. They are listed as follows and summarized in Fig. 10.1.

A. *Technical Risk*: These risks include the following:

- Lack of thorough site investigation.
- Incomplete design.
- Incorrect specification details.
- Uncertainty about material availability.

B. *Logistical Risk*: Examples of these are as follows:

- Lack of transportation facility.
- Lack of maintenance personnel.
- Lack of equipment availability.
- Lack of operator availability.

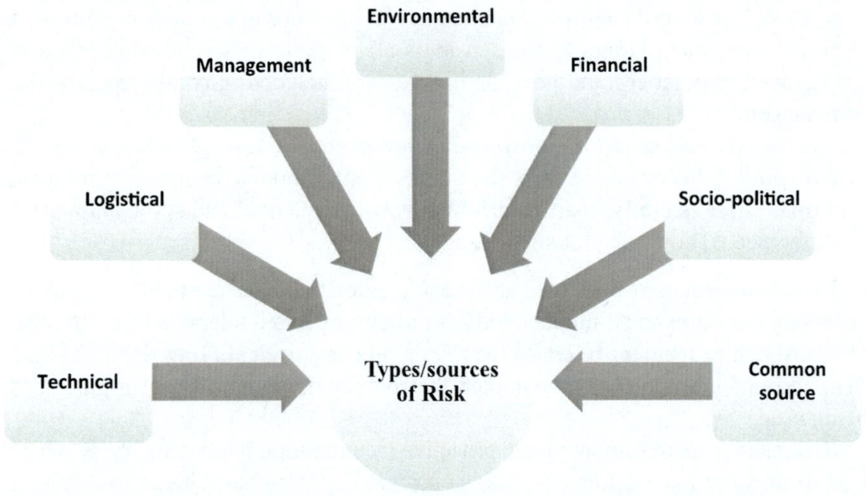

Fig. 10.1 Types/sources of risks involved in construction

C. *Management-Related Risk*: These include the following:

- Unstable management.
- Uncertain productivity.
- Problematic industrial relation.

D. *Environment Risk*: These include the following:

- Fluctuations in foreign exchange.
- Payment delay.
- Inflation.
- Local taxes.

E. *Financial Risk*: These include the following:

- Customs and import restriction.
- Constraints on availability of expatriate staff.
- Insistence on the use of local firms and agents.
- Constraints on employment of expatriate staff.

F. *Socio-Political Risk*: These include the following:

- Natural disaster.
- Weather implication.
- Design error and omission.

G. *Common Source Risk*: These include the following:

- Change in project requirement.
- Inadequate in responsibilities and staff.
- Inefficiency of skilled staff.
- Unfamiliarity with local conditions.
- Unavoidable accident.
- Inadequate in contractor experience.

Steps in Risk Management Process

Berg (2010) identifies the following as steps involved in the risk management process:

1. The first step is to define the risk environment, goal, and context.
2. The second step is to identify the risk (risk identification).
3. The third step is to assess the risk (risk assessment). Here, the possible impact and magnitude of the risk are quantified. Also, the profitability and the cost implications are evaluated, and finally, the risks are prioritized and ranked accordingly.
4. The fourth step is to respond to the already assessed risk (risk response). This step checks if the risk is acceptable or not. If it is acceptable, the risk is accepted,

and if it is not acceptable, there would be a further enquiry as to whether the risk can be mitigated or not. If the risk can be mitigated, then there would be an action to reduce the risk. If it cannot be mitigated, it should be avoided, transferred, or shared among the parties involved.

5. The fifth step is to control all measures put in place to mitigate risk (control activities). These activities include policies or actions taken by the management to ensure the mitigation of already assessed risks, which can be preventive or corrective.
6. The sixth step is to get all stakeholders on the project adequately informed (information and communication). This is a very important tool for passing across information and understanding the risk management process.
7. The last step is to ensure that the desired outcomes of the risk response are carried out (risk monitoring).

Framework for Risk Management in Canada

One of the first risk management research papers published by the public sector in Canada was titled "Best Practices in Risk Management: Private and Public Sectors Internationally" on April 27, 1999. This research was conducted by Klynveld Peat Marwick Goerdeler (KPMG) for the Treasury Board of Canada Secretariat. Nonetheless, the *Integrated Risk Management Framework (IRMF)* appears to be the most significant of all publications by the Treasury Board of Canada Secretariat. It was written by Robillard Hill in the year 2001. The publication aimed to improve risk management practices in the public sector. The author contends that an IRMF may help achieve this by supporting the public sector's four management commitments: emphasis on citizens, values, performance, and financial prudence.

Creating a holistic risk management framework allows for more participation by the citizens and the construction professionals in order to arrive at a more desirable decision-making. Furthermore, it promotes essential public service principles such as honesty, integrity, and credibility at all levels, resulting in improved outcomes by proactively managing risk. Robillard (2001) argues that holistic risk management promotes a whole-of-government approach (WGA) based on logical prioritization and financial prudence principles.

In addition, Stephen Hill's work *"A Primer on Risk Management in the Public Service,"* published from the University of Calgary in 2001, is a notable and extensively used work in Canada. Hill's work is a brief introductory piece that is not intended to be a thorough study or treatment of risk management, but rather to provide the most fundamental concept of good risk management and challenges that might be faced in implementing risk management in the decision-making process of the public sector. System like this includes the following:

1. Identification of risks and opportunities:
   - Sources of risk.
   - Sources of opportunity.

2. Effective management of risks and opportunities:
   - Assessing and altering risk tolerance.
   - Assessing risks and opportunities.
   - Managing risk.
   - Managing opportunity.
   - Monitoring and using a management control system to examine the strategy, costs and benefits, structure, processes, and risk tolerance.
3. Effective risk and opportunity assessment using return on investment (ROI) in conjunction with real options analysis and/or scenario analysis.

Challenges of Risk Management Practices in Canada

Risk is an essential component of any strategic planning process, as expatiated throughout the course of this chapter. Despite the fact that risk management systems have been established at the national level in Canada, as well as in specific provinces such as British Columbia, there is no clear strategic approach (except the framework put together here) to risk management in the public sector. Finally, as obvious in the global trend, the ability of municipal and provincial governments to create a workable system for risk management is critical in managing the services provided to the people.

Risks that occur within the public sector are significantly greater than risks in the private sector. These risks can come either in the form of financial or capital projects. The Treasury's Orange Book (2020) on risk management provides the local/municipal government in the United Kingdom with a policy guide on managing risk. It recommended that public agencies should consider these three major forms of risk:

1. External risks.
2. Operational risk.
3. Risk related to organizational change.

External Risk comprises political risk, economic risk, socio-cultural risk, legal/regulatory risk, and environmental risk.

Operational Risk is a risk related to the delivery of goods and services. This sort of risk causes losses due to ineffective internal organizational processes and risk management skills.

Risk Related to Organizational Change refers to all practices and acts that exceed present organizational capabilities (The Treasury Orange Book, 2004). These risks are summarized in Fig. 10.2 to facilitate easy comprehension. While the United Kingdom has developed a clear approach to risk management, the same cannot be said for Canada at all tiers of government. Hence, the public sector in Canada needs to develop a risk management approach to address all of the aforementioned forms of risk.

Benefits of Risk Management in Canada

Numerous benefits to risk management implementation were cited in a 1999 KPMG research report published for the Treasury Board of Canada, some of which are as follows:

1. Organizational objectives are met.
2. Improved focus on key business objectives. It also helps managers to direct their resources toward the most important objectives. Problems are not addressed with re-directed resources. As a result, shareholders and ministers have more confidence in the organization. An effective risk strategy is to take action to prevent and limit loss rather than to clean up after the fact.
3. A shift in culture that encourages open dialog about risks and potentially harmful information. Mistakes are tolerated in the new culture, but errors are not concealed. In addition, the culture places considerable emphasis on learning from mistakes.
4. Better operational and financial management by ensuring that risks are adequately evaluated while making decisions. Improvement in operational management will bring about effectiveness and efficiency in service delivery. Managers will have more time to respond and take action if they anticipate problems. On the other hand, the organization will be able to meet its service commitments.
5. Enhancing the planning process and providing a tool to assist management in identifying opportunities.
6. In the short run, it increases management accountability. Also, it increased overall management capabilities in the long run.
7. Increase in value (comment from the private sector) (KPMG, 1999).

Fig. 10.2 Problems with risk management practices in the construction

Conclusion

Risk assessment is a critical component of risk management, which includes risk identification, analysis, and evaluation. The qualitative risk assessment methodology simplifies risk assessment by utilizing three rating methods: frequency and probability ratings, consequence ratings, and risk matrices. The quantitative risk assessment methodology assures precise numerical values for risk by utilizing likelihood ratings, consequence ratings, scoring methods, and ordinal scales. Even though the country's Treasury Board (through the government) has developed a framework toward risk management practice, the development of a functional and applicable framework for the public sector will help the country's quest to fully implement risk management practice into the country's construction.

References

Apostolakis, G. E. (2004). How useful is quantitative risk assessment? *Risk Analysis, 24*(3), 515–520. https://doi.org/10.1111/j.0272-4332.2004.00455.x

Berg, H. P. (2010). *Risk management: Procedures, methods and experiences*. Retrieved from https://www.scirp.org

Betts, M., Robinson, G., Blaik, N., Burton, C., & Godden, D. (2011). *Global construction 2020: A global forecast for the construction industry over the next decade*. Global Construction Perspectives and Oxford Economics.

Bird, F. E., & Germain, G. L. (1996). *Practical loss control leadership*. Retrieved from https://www.googgle.com/practical-loss-control-leadership

Budescu, D. V., Broomell, S., & Por, H. H. (2009). Improving communication of uncertainty in the reports of the intergovernmental panel on climate change. *Psychological Science, 20*(3), 299–308. https://doi.org/10.1111/j.1467-9280.2009.02284.x

Cox, L. A. (2008). What's wrong with Risk Matrices? *Risk Analysis, 28*(2), 497–512. https://doi.org/10.1111/j.1539-6924.2008.01030.x

Cox, L. A., Babayev, D., & Huber, W. (2005). Some limitations of qualitative risk rating systems. *An International Journal on Risk Analysis, 25*(3), 651–622.

Genik, L., & Chouinard, P. (2012). *DRDC support to Emergency Management British Columbia's (EMBC) Hazard Risk Vulnerability Analysis (HRVA) and Critical Infrastructure (CI) programs: Problem formulation and solution strategy*, Defense Research and Development Canada DRDC Technical Memorandum DRDC CSS TM 2012–01, October 2012, Canada.

Grant, R. M., & Venzin, M. (2009). Strategic and organisational challenges of internationalisation in financial services. *Long Range Planning, 42*(5–6), 561–587.

Hartley, M. (2018). *The use of qualitative risk analysis methods to facilitate decision making in the management of health and welfare in wildlife*. Retrieved from https://www.chesterrep.openrepository.com

Hill, S. (2001). *A primer on risk management in the public service*. University of Calgary.

Hubbard, D., & Evans, D. (2010). Problems with scoring methods and ordinal scales in risk assessment. *Journal of Research and Development, 54*(3), 1–10.

Human Resources and Skills Development Canada. (2013). *Human resources and skills development Canada: Reports on plans and priorities*. Retrieved from https://www.canada.ca

Industry Canada. (2012). *Gross domestic product construction*. Canadian Industry Statistics. Retrieved from http://www.ic.gc.ca/cis-sic/cis-sic.nsf/IDE/cis-sic23vlae.html

International Electro-Technical Commission (IEC). (2018). *Risk assessment standards*. Retrieved from https://www.webstore.iec.ch

International Standard Organisation (ISO). (2022). *Risk management*. Retrieved from https://www.iso.org

Kaplan, S., & Garrick, B. J. (1981). On the quantitative definition of risk. *Risk Analysis, 1*(1), 11–27.

Klynveld Peat Marwick Goerdeler (KPMG). (1999). *Best practices in risk management: Private and public sectors internationally: KPMG study prepared for treasury board of Canada secretariat*. Retrieved from https://www.researchgate.net

McGee, M. W. (2005). *Measuring the payoff of strategic risk management*. Retrieved from https://www.thefreelibrary.com/Measuring+the+payoffs+of+strategic+risk+management.a0142620212

Paté-Cornell, M. E. (1996). Uncertainties in risk analysis: Six levels of treatment. *Reliability Engineering and System Safety, 54*(2), 95–111.

Robillard, L. (2001). *Presidents Message, Treasury Board of Canada Secretariat. Integrated Risk Management Framework*. Retrieved from http://www.tbs-sct.canada.ca

Society for Risk Analysis. (2022). *A community for risk science-society for risk analysis*. Retrieved from https://www.sra.org

The Treasury Orange Book. (2004). *Management of risk – Principles and concepts*. Retrieved from https://www.hm-treasury.gov.uk

Weaver, P. (2008). *The meaning of risk in an uncertain world*. Retrieved from https://www.pmi.org

Wilson, L., & McCutcheon, D. (2003). *Industrial safety and risk management*. University of Alberta Press, Edmonton.

Chapter 11
Risk Management Practices in Sweden

Abstract Infrastructural development in any country would certainly enhance the economic growth of such a country and generate job opportunities. Through different practices and concepts, the construction industry in Sweden has improved tremendously over the years. Several attempts have been made in Sweden to address the issue of risk management using various methodologies. This chapter assessed the Work Environment Act (AML) and other construction bodies or regulations that are part of Swedish law that ensures adequate risk management in construction contracts with the aim of bridging knowledge gap. Other sections of the chapter discussed the history of risk management in the country, factors that affect effective implement of the practice, and other related topics. The chapter concluded that more still needs to be done in order to move the country's construction toward that which is risk-free through identified risk management frameworks used in several other developed countries of the world.

Keywords Construction risk · Project delivery · Project risk · Risk framework · Risk management · Risk valuation

Introduction

A risk can be defined as the probability or proclivity of an event or problem occurring in relation to the success of a project, achieving the project's desired outcome, or an uncertain future occurrence with a probability greater than 0% but less than 100% that would have an effect on the project's objectives (cost, quality, and time) (Mark et al., 2004). Additionally, the impact of this future occurrence must be unpredictable or unforeseen; it has been postulated that risks can be mitigated to upturn their resulting effect on project success even if their occurrence is unavoidable during the life of a project (Chia, 2006). There are various sources of risk. These include intrinsic uncertainties, institutional inadequacies, and company-related issues such as technological implementation, economic policy, weather and

A. E. Oke et al., *Risk Management Practices in Construction*,
https://doi.org/10.1007/978-3-031-35557-8_11

climatic condition, inflation, market competition, and bidding process (Karimi et al., 2011).

Project risk management has been practiced since the mid-1980s, and it is currently considered one of the nine core specialist areas of the project management institute's body of knowledge (Tuysuz, 2006). Risk can be mitigated by observing, monitoring, managing, analyzing, and applying managerial skills and resources through synchronized socioeconomic and political stimulation to minimize the eventual of the unforeseen event occurring (Douglas, 2009). Effective and efficient practice of risk management would lead to enormous benefits with respect to the success of a project; such benefits include identification of the favorable alternative course of action, adequate site investigation, proper project planning and budgeting, adequate price forecast, proper feasibility study, improved technological implementation, and proper and adequate estimates (Bannerman, 2008).

The ability to achieve a sustainable construction project is expanding, and higher standards are expected, which has led to the development of technology, improved material, and advanced technique application. It is common knowledge that every new technology has its constraints and risks; the desire to achieve expected results through utilizing these modern technological techniques and materials has made the importance of project risk management increase and its awareness more profound. Infrastructural development in any country would certainly enhance the economic growth of such a country and generate job avenues. The cost of undertaking such a project is enormous. It delivers on time, and to standard, project risk management practices must be involved so as to prevent huge monetary losses (Bannerman, 2008).

Risk is critical to completing a project within the stipulated time period and cost budget. The construction industry is frequently regarded as a high-risk venture owing to its complex and strategic nature. This venture involves a large number of stakeholders and internal and external influences, all of which pose a significant risk. Unfortunately, when compared to other sectors, the construction industry has a bad reputation for risk analysis and management (Laryea & Hughes, 2008). No construction project, however, is risk-free, yet these risks can be controlled, mitigated, shared, transferred, or accepted (Latham, 1994).

About Project Management Institute (PMI) (2008), the risk is a multi-facet concept; the processes involved in identifying, analyzing, and responding to project risk are referred to as project risk management. In the construction industry, it refers to the probability of a given occurrence or series of events occurring throughout the course of construction. Construction involves a vast number of factors, and establishing cause and effect, dependence, and correlations is frequently difficult.

Risk Management in the Swedish Construction Sector

In the past, the Swedish construction industry was known to struggle with challenges such as the increase in construction costs, delays in the delivery of projects, and poor-quality standards. To address these concerns, the federal government formed a commission, which issued the "Skärpning Gubbar" Report 2002 in Simu (2007). According to the commission's reports, despite continued quality management and internal quality surveillance operations, significant gains were not being made. Risk management basically involves the ability to think ahead and plan for unforeseen events and scenario that would affect a project success, which is embedded in the duties of a quality manager of any organization. The Swedish system has been designed to prioritize medium- and large-scale initiatives, with little or no attention paid to smaller projects. However, small projects are an integral part of the industry, constituting about 83% of the total number of projects undertaken in a fiscal year.

The Swedish construction industry has been impeded by poor performance and sub-standard project quality. With respect to these numerous faults, the Swedish government established the Building Commission to identify issues and propose solutions to improve project delivery effectiveness and reduce delays and cost overruns. The commission-found major faults related to a construction project are a result of inadequate risk management, which leads to increased project delivery costs, delay, and poor quality of work. The Commission discovered that quality management systems in accordance with ISO standard 9000 were insufficient to address the sector's fundamental shortcomings (Simu, 2007).

Risks are uncertainties that occur in various ways and are linked to practically all aspects and stages of a construction project. The objectives of every project are linked to its overall cost, the quality standard of that project, its functional rating, and satisfaction derived by clients (Hillson, 2004). Organizations have different understandings and definition of risk management, which results in various techniques applied in the management of risks. The majority of managers in the Swedish construction industry focus on-site management (Hillson, 2004), which is also known as probable alternative to project risk management and safety risk management. They tend to use techniques connected to various management systems at later phases.

Risk management is a component of the management system with several similarities to quality management, environment, or work environment management system. The core values shared among these branches of management include managerial commitment, employee commitment, quality of work done, continuous improvement, and adequate decision-making ability. Risk management is an important aspect of project management, and with the current concentration given to risk management, it should be regarded as an improvement on the quality management system used by most companies. The majority of the projects in the Swedish construction sector are under 15 MSEK (approximately €1.65 million). In Hultén (2004), 83% of all projects ranged from 1 to 15 MSEK in the year 2003. The comparable figure was 68% at NCC (data from NCC's Internal Economical Control System from 2003). The distribution of projects in each sector, sorted by monetary value, indicates that the bulk of construction projects completed in Sweden are on a small scale. In the Swedish construction sector, projects worth up to 50 MSEK account for half of the industry's turnover; this is not to say that projects in the span of 1–15 MSEK are not important, but they are still countable. Most of the large projects undergone in Sweden employ the use of risk management technique, which is because organizations that engage in this construction have the money and time to retain the services of risk management professionals so as to achieve the desired objective of the project (Hellström, 2006). Large-scale projects undergo proper planning, and the professionals involved have meaningful input at the initial stage to highlight the most favorable risk, analyze its outcome, and further proffer meaningful solutions.

The small-scale projects, on the other hand, have limited budgets and time constraints, which most likely interfere with proper planning even before execution. This eventually leads to potential risks not being identified at an early phase of the project, which finally results in increase in cost and an extended completion period.

The Swedish construction industry has a standard agreement for the contract. It consists of different agreements subject to the form of a contract. Describing the contract, AB04 is used for general contracts, while ABT94 is utilized for build and design contracts (Berg, 2010). When preparing these general agreements, the Construction Contracts Committee (Byggandets Kontraktskommitté, BKK) is in the duty of representing numerous parties in the Swedish construction industry. In a general contract, responsibilities are regulated between the various actors to the project (contractor and client), and parties charged with the function of preparing designs and undertaking the construction work are stated clearly with the ultimate aim of achieving the best results.

Under this general contract, the contractor is fully accountable for the construction work under the general contract, and it is the client's responsibility to make any adjustments or modifications to the design. This enables the contractor to focus on the major task at hand and plan toward all unforeseen event and their likely outcomes. The contractors are often compensated for delays in work due to design modifications. Clients, on the other hand, are aware of these laws; therefore, any anomalies with respect to the contractual agreement tendered by the contractors are seen as a means to request unnecessary claims and attract extra cost. Since small-scale projects are most likely subjected to design modification, the need to plan for inherent uncertainties should be proper and adequate. Parties to this agreement are strongly advised not to deviate from these agreements.

Today, insurance does not cover contracts that depart from AB's general agreements (Hellström, 2006). As a result, it is critical that the people who sign the contracts understand everything. In Hellstrom (2006), risk insurance in the Swedish Construction industry for large projects is between 100,000 and 200,000 SEK (10,000–20,000€), which is a tiny portion of the entire project cost. For small-scale projects, the risk insurance is between 25,000 and 100,000 SEK (2500–10,000€), which is a larger proportion of the total contract sum. This makes it difficult for small-scale projects to prepare extensively against the risk associated with the construction work due to financial constraints, among other reasons.

The risks associated with the workplace are strictly governed by laws and regulations. The Work Environment Act (AML), passed by the Riksdag (Sweden's parliament), governs the Swedish Work Environment Authority (Government Offices of Sweden, 2015). The Work Environment Authority then issues detailed regulations on the subject via provisions and general recommendations. It is stated clearly within the regulation the description for what activities is to be undertaken and who is to undertake such activities. It is customary for the client to delegate the work to be done on-site to the contractor, who in turn makes necessary provisions to militate against various risk factors. The main contractor then drafts the AML-required paperwork, such as risk studies for workplace risks and a work environment management plan.

Factors that Affect Effective Risk Management in Sweden

Osipova and Atkin (2008) conducted a study on nine construction projects in Sweden, which include road, residential, and other facilities. Risk management was implemented in each of these projects, but it was not done systematically, thus resulting in cost overruns in some of them. The findings of this study explained some important factors influencing effective risk management in Sweden. They include the following.

Understanding of Risk Management

There is a need for project participants to understand risk management. According to the quoted study above, most project participants have a basic understanding of risk management as a result of their experience in the construction industry. Some organizations also train their employees on risk management; the only challenge is implementing it consistently. The researchers went further to identify some of the risk management tools that project actors are familiar with, such as checklists, and to brainstorm for risk identification, probability-consequence for risk assessment, and risk transfers for risk response.

Participation at Different Levels of the Project

The construction process is systematic, with multiple parties involved in various stages of the project, each focusing on their specific aspect of the task rather than the overall project. As a result, identified threats are communicated less effectively, and knowledge is lost between stages.

In the identified study, there is little participation during the planning stage, and contractors are not even participating at this stage of the project. Furthermore, it was recommended that the presence of a contractor at this stage is critical in helping the parties concerned to select the best technical approach, understand the cost, and anticipate future challenges.

Risk Identification, Assessment, and Response

Systematic investigation of probable and possible risks in the project is crucial for successful risk management. The term "systematic" refers to risk identification, assessment, and response at every level of the project, and the results of the processes are communicated to project stakeholders.

- *Risk Identification*

This is considered to be the first step in militating against risk. This step involves the use of techniques and tools to brainstorm and break down the inherent risk (Maytorena et al., 2005). In the quest to identify risks associated with a certain project, it is necessary to first identify the different sources of these risks and their impacts on the project. This would enable one to set priorities right and proffer suitable solutions (Flanagan & Norman, 1993).

- *Risk Assessment*

This process entails the application of tools to determine the probability of an unforeseen event occurring and its predicted impact, allowing risk management to categorize such events into a spectrum of most likely to least likely (Grey, 1995).

- *Risk Response*

Risk response is the process where steps are taken to mitigate the effects of risks assessed. The four levels that are usually used to grade responses are risk retention, risk reduction, risk transfer, and risk avoidance (Flanagan & Norman, 1993).

Risk Transfer and Communication of Risks

Risks are typically shared among project participants. Some risks are intrinsically assigned to the client, such as market and environmental risks; the client transfers some risks to the contractor because they feel they have the necessary technical competence to manage the risk.

Risks are either positively or negatively transferred. The positive approach entails carrying out risk identification and ensuring that the party responsible for risk management is fully aware of the risk prior to the start of the project. The negative approach involves the client omitting some information prior to the start of the project and the risk later appearing during the construction phase that must be addressed.

The findings of Osipova and Atkin (2008) further provided two viewpoints on the negative risk transfer approach. The first contended that some clients might choose not to disclose all known risks during the procurement stage in order to reduce the value of the bid and simply assume that the risk can be managed by the contractor. The second contended that both the client and the contractor are unaware of the potential risk and must reach an agreement on how to address it. Therefore, accurate and complete risk identification is required during the planning stage, and risks are communicated at a detailed level to limit the chances of their occurrence throughout the course of construction, as well as their impact.

Joint Risk Management

Joint risk management is a technique that emphasizes the need for project participants to work together to manage risks. This is essential to prevent the contractor from being forced to manage the risk in a project, which is frequently the case in most projects.

The following are some critical factors that must be considered to accomplish effective collaborative risk management:

1. The participation of all project stakeholders in risk discussions.
2. The effectiveness and openness of communication and information exchange.
3. The ability of project participants to highlight issues as they arise, to ask probing questions, and to work without concern for prestige.
4. The commitment, motivation, and integrity of each participant involved.
5. The trustworthiness.
6. The mutual respect for each other's positions and competency.
7. The equitable allocation of opportunities.

Conclusion

This chapter showed that most capital projects executed in Sweden use risk management techniques, which is because organizations involved in this construction have the money and time to retain the services of risk management professionals. Capital project undergoes proper planning hence reducing risk owing to the involvement of different construction professionals in executing project planning to execution. On the other hand, small-scale projects have limited budgets and time constraints, which most likely would interfere with proper planning, eventually leading to potential risks not being identified at an early phase of the project. This affects the outcome of the construction but can be managed by following strict conditions (rules and regulations) associated with the Swedish construction industry.

References

Bannerman, P. L. (2008). Risk and risk management in software projects: A reassessment. *Journal of Systems and Software, 81*(12), 2118–2133.

Berg, H. P. (2010). *Risk management: Procedures, methods and experiences*. Retrieved from https://www.scirp.org

Chia, E. S. (2006). *Risk assessment framework for project management*. Retrieved from https://www.ieeexplore.ieee.org

Douglas, H. (2009). *The failure of risk management: Why it's broken and how to fix it*. Wiley.

Flanagan, R., & Norman, G. (1993). *Risk management and construction*. Blackwell Scientific Publications.

Government Office of Sweden. (2015). *The Swedish work environment authority activities 2015*. Retrieved from https://www.av.se

Grey, S. (1995). *Risk assessment for project management*. Wiley.

Hellström, M. (2006). *Risk manager*. NCC Group. Retrieved from https://www.google.com

Hillson, D. (2004). *Effective opportunity management for projects – Exploiting positive risk*. CRC Press.

Hultén, V. (2004). *Statistics from Sveriges Byggindustrier.* Stockholm.

Karimi, A., Mousavi, N., Mousavi, S. F., & Hosseini, S. (2011). Risk assessment model selection in construction industry. *Expert Systems with Application, 38*(8), 9105–9111.

Latham, M. (1994). Constructing the team. In *A final report of the joint review of procurement and contractual arrangements in the United Kingdom construction industry*. Crown Publication.

Laryea, S., & Hughes, W. (2008). How contractors price risk in bids: Theory and practice. *Construction Management and Economics, 26*(9), 911–924.

Mark, W., Cohen, P. E., & Glen, R. P. (2004). *Project risk identification and management*. Retrieved from https://www.proquest.com/openview

Maytorena, E., Winch, G. M., & Kiely, T. (2005). Construction risk identification. In *Proceedings of the 11th joint CIB international symposium- combining forces advancing facilities management and construction through innovation*, 13–16 June, Helsinki.

Osipova, E., & Atkin, B. (2008). *Risk management in construction projects: A comparative study of the different procurement options in Sweden*. Building Resilience.

Project Management Institute (PMI). (2008). *A guide to the Project Management Body of Knowledge (PMBOK guide)*. Project Management Institute Incorporation.

Simu, K. (2007). *Risk management on small projects*. Retrieved from https://www.diva-portal.org

Tuysuz, K. (2006). Project risk evaluation using a fuzzy analytic hierarchy process: An application to information technology projects. *International Journal of Intelligent Systems, 21*(16), 559–584.

Chapter 12
Risk Management Practices in the United States of America

Abstract The study of risk management (RM) began in the aftermath of World War II. It began with market insurance and then moved to other industries, such as finance, stock exchange, and construction projects, before finally becoming widespread. In its most basic form, risk management seeks to safeguard individuals and businesses against financial losses that may result from accidents. The US construction industry is one of the major pioneers of risk management concepts developed over the years. This concept has been reviewed over time in order to meet the present demand for sustainable buildings. Through different sections, by identifying the sources of risk, the purpose of risk management, and other related practices, the study gave an exclusive overview of risk management practices to achieve enhanced project delivery.

Keywords Project risk · Project success · Risk identification · Risk management · Risk properties · Sustainable building

Introduction

Risk is common in every industry. Even in different formats across several phases, its effects are felt either positively or negatively along the line. To find a solution to this, risk management was introduced to enlighten, direct, and discuss risk properties across business operations, industries, supply chains, products and services, etc.

The beginning of modern-day risk management can be traced back to 1955–1964, according to scholars such as Crockford, 1982; Williams & Heins, 1995.

In the early days, according to Snider (1956), there was no literature on risk management, and no tertiary institutions (universities) offered courses in risk management at the time of observation. Regarding this, the first book was believed to have been published by Mehr and Hedges in 1963, and Williams and Hems released the second in 1964. The subject matter presented at the time was pure risk management, which did not include financial risk assessment/analysis of companies.

A. E. Oke et al., *Risk Management Practices in Construction*,
https://doi.org/10.1007/978-3-031-35557-8_12

Later in the years, engineers devised a model for technological risk management in operational risk, which partially compensated for technology losses. Progressively, operational risk was embraced and employed by financial organizations such as banks, insurance companies, and stock exchanges.

Risk management is a contemporary corporate practice that has evolved in recent years. After 1955, modern-day risk management became a reality. During the 1970s, the notion of risk management underwent significant evolution. A notable development has been the expansion of risk management beyond market insurance coverage, which is now considered a competitive protection instrument that complements a variety of other risk management operations (Sousa et al., 2012).

Typically, construction projects are carried out in a complicated and ever-changing environment, which creates situations characterized by a high level of uncertainty and risk and are constrained by strict time limitations (Schieg, 2006). Historically, the construction industry has been and continues to be a very dynamic sector that has undergone and will continue to undergo changes due to demands, economics, utility, policy, etc. This industry is mostly pushed by the private sector, resulting in a significant increase in the presence of real estate across nations. As a result of these factors and more, the construction sector is susceptible to a wide range of technical and business risks, many of which are more severe than those encountered in other traditional industries. This situation, therefore, necessitates the need for risk assessment to be fully implemented in construction to help fill lapses overlooked or not considered right from the onset of project planning. When it comes to project execution, risk assessment is a strategy that helps to detect potential risks and manage them appropriately with appropriate treatment.

The United States of America in the Construction Industry

The construction industry accounts are crucial to the development of any nation. In the United States, the construction industry contributes the largest share of the country's gross domestic product (GDP) (Arditi & Mochtar, 2000). From 1995 to mid-2000, construction expenditure accounted for around 7–8% of yearly economic production in the United States, peaking at about 9% of GDP during the housing boom in 2006 before falling to under 5% in 2010. Similarly, construction employment varied from approximately 4.6 million during the 1991 crisis to a high of 7.7 million in 2006 before falling to approximately 5.5 million in early 2011 (Kirchhoff, 2011). In 2020, the industry was expected to expand by 4.5% in real terms after a $1.5 trillion spending package was signed by the president of the country (Emam et al., 2015).

The construction industry in the United States is categorized into various sectors: residential, commercial, industrial, institutional, energy and utilities, and infrastructure. As big as the industry is, it is not immune to risk. Relating the practice to the United States (US), Kangari (1995) explained the early reaction of big US construction companies toward risk management and evaluated how the contractors implement construction risk management using a survey of the top 100 contractors. According to the findings of the study, contractors have become more eager in recent years to bear risks associated with actual and legal problems, as well as share risk with the property owner. This has helped parties concerned with project delivery to pay more attention to details, as failure may lead to damages on all fronts.

Now, the industry has grown massively, and the implementation of risk-related procedures has helped the industry achieve its robust nature. Coupled with other digitally enabled concepts and practices put in place, the country's industry can be said to be at the forefront of project enhancement, project delivery, and other improvements experienced over the years in both developed and developing nations of the world.

Risk Concepts in the US Construction

Risk, in simple terms, can be defined as the likelihood of certain event or combination of events occurring during the entire construction process (Mehr & Hedges, 1963). Also, it can be seen as a lack of certainty as to what the outcome or impact might be at the planning or decision phase. There is a possibility that results will be better or worse than expected because of the uncertainty connected with outcome.

In any country, construction is seen as a significant industry. This is attributed to the qualities of its products, its contribution to the GDP, and improvement that comes in the form of enhanced quality or standard of living. For example, in the United States, investment in design and construction services currently amounts to approximately one trillion dollars per year, which amounts to approximately 10% of the country's GDP on average. This figure is replicated several times throughout the world (multiplier effect) (Emam et al., 2015).

There are several definitions of risk, along with classifications used for different purposes. In construction projects, the risk is divided into two categories: external risks and internal risks. Internal risks are risks that occur as a result of poor performance by the employee, accidents in the production facilities, operational efficiency, etc., while external risks include political risk, financial risk, and market risk.

Objectives of Risk Management

The following are some of the objectives of risk management in the construction industry:

1. To actively manage risk to enhance project success.
2. To make risk proportional to project size and complexity.
3. To integrate risk management into the present project delivery process.
4. To include all functional units in the process, etc. (Emam et al., 2015).

Purpose for Studying Risk Management

Construction has historically been a high-risk endeavor for a variety of reasons, and studying and embracing risk management practice will help to reduce potential risks from onset of project planning to execution. Studying risk management will aid the following:

1. The pressure to save time and money.
2. The pressure to deliver a high-quality job.
3. In generating a high return on investment (ROI).
4. In tracking and keeping records of projects (completed or uncompleted).
5. By making project delivery to be on schedule and within budget.
6. In reducing or eliminating dispute and litigation.
7. In improving profit and market value.
8. In project delivery beneficial to both the contractor and the client, etc. (Emam et al., 2015).

Sources of Risk

The number of potential sources of risk that may affect a project is virtually limitless in theory, but in practice, the number of statistically significant sources of risk is quite modest.

For the vast majority of construction projects, the following are the most important sources of risk (Schieg, 2006).

Characteristics of the Client

Depending on the situation, the client may be a public entity, a commercial corporation, or an individual. Consider the track records and attitudes of the client regarding professional services, as well as the availability and adequacy of funds for the project, among other considerations.

The Nature of the Project

The location of the project, budget, and political profile in the community where it is located is considered. Also, the laws governing the implementation of such projects are all important considerations. The project's simplicity or complexity is another point to be considered.

The Usage of Consultants for Professional Services

For the project, the availability of skilled and experienced experts is important to consider. Also, the consultants' professional reputation should be within standard and required education.

Contractor and Mode of Project Delivery

Whether the construction contract will be awarded through open bidding or selective bidding, whether there will be a single primary contractor with subcontractors providing services, whether or not the contract documents will be completed right before the commencement of the construction project, and whether the design expert will be properly compensated and safeguarded are important considerations and questions to be answered.

Professionals' Abilities and Previous Experience

Check to see if all of the specialists involved are licensed, if the firm has previous experience with the type of project, and if the firm has enough staff and consultants on hand to provide the services on time and within the project budget.

Types of Contracts for Design and Construction

The contract between the design expert and the client is prepared using a standard industry form, for example, the American Institute of Architects (AIA). The client will engage a number of prime design professionals to provide services in a variety of disciplines. The methods of remunerating specialists, the structure of the building contract, and the basic terms and conditions should be discussed.

Identifying Project Risks

The project risk manager is the officer put in place to convene risk project meetings and identify and assess risks. There is always the challenge of differentiating between risk sources, genuine risks, and risk impacts when it comes to risk identification.

Causes Some specific events or situations exist in the project or its environment, such as follows:

- The need to use unproven method.
- Scarcity of qualified employees.
- The fact that the company has not previously completed a project of similar nature.

Genuine Risks These are uncertainties that, if they happen, would affect the project objectives in either a negative or positive way. The negative is referred to as a threat, while the positive is known as an opportunity. The uncertainties must be treated in a proactive way.

Effects These are unanticipated changes or deviations from the project objectives that could occur as a result of the risk, e.g., failing to meet the agreed quality, exceeding authorized budget, and not meeting the milestone time targeted.

How Risks Are Identified

During risk identification, any of the following or a combination of them can be used:

- Brainstorming by the Project Risk Management Team (PRMT).
- Challenging assumptions (e.g., using reinforced concrete instead of timber floors).
- Looking for newness.
- Experience from similar projects.
- Consultation with people who have had previous experience with similar projects.
- Experiential learning from stakeholders or other individuals within the firm/organization (Jayasudha & Vidivelli, 2016).

Project Risk Management Process

Construction risk management is designed to plan, monitor, and control measures required to avoid risk exposure. Project risk management will enhance the achievement of project objectives which are as follows:

1. There will be no cost overrun (to cost).
2. there will be no time overrun (timelines).
3. Quality will not be compromised.

In essence, there will be value for money, the project will be executed within budget, and there will be a balance (logical) distribution of resources in all areas of the project.

The following steps are included in the project risk management process (Sousa et al., 2012):

1. Risk management planning.
2. Risk identification.
3. Risk analysis—both quantitative and qualitative risk analysis.
4. Risk response.
5. Risk monitoring.
6. Risk communication.

Risk Management Planning (RMP)

This is the first step to be taken in the implementation of a risk management plan (RMP). It specifies the extent to which risk management will be carried out, that is, the development of the project roadmap. Risk management planning is not required for all projects; the need for one is determined by the size and complexity of the project, as well as the amount of risk management effort required (Jayasudha & Vidivelli, 2016).

The RMP determines the frequency of meetings and the location and frequency of risk register updates. In addition to a list of the risk management team members who will be involved, it includes a budget for the risk management activities. The RMP must be completed as soon as possible due to its critical role in successfully implementing other procedures.

Steps to Risk Management Plan

- Determine the scalability level of the project.
- Put in place a risk register.
- Determine the frequency with which project meetings will be held.
- Make a decision on who will be a member of the Project Risk Management Team (PRMT).
- Include cost estimate for external consultants in your budget.
- Obtain the approvals that are required for the RMP.

First Project Risk Management Meeting Action Plan

The very first meeting of PRM is extremely crucial; all members must be present because salient and very important decisions will be made during this meeting. It is at this meeting that the project manager should inform the team of the following:

- The significance and objectives of the project risk management process will be highlighted.
- The duties and responsibilities will be shared among the participants.
- Risk register will be put in place.
- The communication channel and accountability will be made known.
- Schedule of risk management activities.
- Codes for calculating time charges for risk management activities.
- The expectation that risk will be managed, documented, and reported on.

Risk Identification

Since it is difficult to completely eliminate risk, it is necessary to assess and deal with all of the risks that have been identified. The identification of risks must be done in a way that is both forward-looking and in synchronization with the progress of the project. As a result, all critical risks must be covered to the greatest extent possible.

Risk identification determines any occurrence that could occur and have an impact on the objectives of the project, as well as the manner in which those events could occur.

At this point, it is necessary to establish a project risk register, which will contain information on the risks and their characteristics. The risks may either be negative or positive, which can also be terminal threats and opportunities.

When the contractors secure the project, the project objectives (scope, time, cost, and quality) are contractually established and cannot be changed. There are fewer opportunities to save money and time (Jayasudha & Vidivelli, 2016).

The Risk Register This is a document that lists all of the risks that could occur. It can be utilized by the project teams to address and document potential risks throughout the life cycle of the project. It is a living document that incorporates significant risks and information on how they are being addressed. This register is included in the project file. The risk register is an extremely important tool because it conveys project risks and assists members of the risk management team (RMT) in understanding the condition of risks as the project progresses from the inception phase to the final completion.

A project risk register should be prepared alongside the cost and schedule of the project. This document must be reviewed and updated at the start of each project phase. A risk register must be used throughout the life cycle of the project, from the initial project document to construction, so as to capture the systematic evolution of project risk.

Resolution of Disputes

There must be a mechanism for resolving disputes among the Project Risk Management Team (PRMT) so that they can achieve success at the end of their assignments. They must put in place a dispute resolution ladder (DRL) that clearly outlines the processes through which disputes will be discussed and ultimately resolved. The project manager (PM) will be the one to facilitate resolution and its elevation.

Risk Analysis

This process is where the risk is described. When describing a risk, the objective is to express it completely and precisely as much as possible. The next stage is to prioritize or rate them and then investigate them to estimate the likelihood of their occurrence and the impact they will have on the project that has been established. Broadly speaking, we have two ways of analyzing risk, i.e., qualitative and quantitative risk analysis (Jayasudha & Vidivelli, 2016).

Qualitative Risk Analysis

This is a method of prioritizing the identified risk for further investigation, which will lead to the type of response that will follow.

The PRMT may revisit the qualitative risk analysis along the life cycle of the project. This is done to check whether the risk mitigation adopted is working. Prioritizing, in this case, simply means rating. The rating will help the team to focus their attention on the ones with higher ratings. The rating will be in three layers.

"High"—a first priority for quick response.
"Medium"—Response as time and resources permit.
"Low"—Response is not necessary at this time.

Quantitative Risk Analysis

This is another method of analyzing risk on a project by means of numerical estimating. This analysis checks whether the project will meet its objectives (time and cost). There is software called "Primavera Risk Analysis software" for simultaneous evaluation.

It is vital to note that this approach of risk analysis is heavily influenced by two significant factors: time (the completion term) and cost. If the mentioned software is used, the result will be a probability distribution of project cost and completion time based on the identified risk.

Any activities or line-item cost element contributing most to the possibility of exceeding time or cost targets can be determined, and then, due attention can be focused on such by performing sensitivity analysis.

The chance of the risk occurring is represented by the two values "Low" and "High" in this context. Another example is the use of three estimates for both cost and time. This consists of the words "Low" (optimistic) and "High" (pessimistic), and the last is "Most Likely" for the cost and time.

Under this quantitative risk analysis, figures are used all along to generate a table, which can further be used to plot graphs at different times in order to have a comparison.

Risk Response

This is the means by which the project risk management team responds to all risks identified on a project. This is done by establishing strategic options and formulating action plans in order to maximize opportunities while maintaining project objectives. Every member of the PRMT must be alive of their responsibilities in handling threats assigned to them (Sousa et al., 2012).

Risk Monitoring

Continuous monitoring must be done by the project team to guarantee that new and changing risks are identified and effectively managed. In addition, the team must ensure that risk response actions are conducted effectively and efficiently to achieve the project objectives and success across the project phases.

Risk monitoring must be done throughout the life of the project for the sake of sustainability. The project team should have their meeting regularly to update the risk register and have project risk reviews.

Risk Communication

Communication is a critical component in the management of project risks. Everybody involved in the project at hand must be adequately informed. Communication must be simple, clear without any ambiguity. Communication and consultation with all stakeholders, sponsors, and the project team are critical components of effective risk management practice (Schieg, 2006).

Conclusion

Effective project risk management must penetrate all project areas, functions, and activities. The ultimate goal must be negotiating risk assets and reducing or eliminating them completely. The key to achieving this is for all parties involved to demonstrate dedication while also acting with professionalism. Also, the parties involved must be highly motivated to do a thorough job because the quality of their work and the success of the project are determined at the end of the process. Managing risks is critical to successfully implementing any project because it demonstrates an understanding of the project's goals, responsibilities, service content, and feasibility. It is expected that the construction professionals in the United States will continue to implement risk-related practices and develop more concepts that will aid construction across all phases.

References

Arditi, D., & Mochtar, K. (2000). Trends in productivity improvement in the US construction industry. *Construction Management and Economics, 18*(1), 15–27. https://doi.org/10.1080/014461900370915

Crockford, G. N. (1982). The bibliography and history of risk management: Some preliminary observations. *The Geneva Papers on Risk and Insurance – Issues and Practice, 7*(2), 169–179.

Emam, H., Farrell, P., & Abdelaal, M. (2015). Causes of delay on large infrastructure projects in Qatar. In *31st annual ARCOM conference,* September 15–17, Lincoln.

Jayasudha, K., & Vidivelli, B. (2016). Analysis of major risks in construction projects. *Journal of Engineering and Applied Sciences, 2*(11), 6943–6950.

Kangari, R. (1995). Risk management perceptions and trends of United States construction. *Journal of Construction Engineering and Management (American Society of Civil Engineers) (ASCE), 121*(2), 201–208.

Kirchhoff, S. M. (2011). *The construction sector in the U.S. economy.* Retrieved from https://www.crs.gov

Mehr, R. I., & Hedges, B. A. (1963). *Risk management in the business enterprise*. Literary Licensing.

Schieg, M. (2006). Risk management in construction project management. *Journal of Business Economics and Management, 7*(2), 77–83.

Snider, H. W. (1956). Reaching professional status: A program for risk management. In: Corporate risk management: Current problems and perspectives. *American Management Association, Insurance Series, 112*(1), 30–35.
Sousa, V., Almeida, N. M., & Dias, L. A. (2012). Risk management framework for the construction industry according to the ISO 31000:2009 standard. *Journal of Risk Analysis and Crisis Response, 2*(4), 261–275.
Williams, A., & Heins, M. H. (1995). *Risk management and insurance*. McGraw-Hill.

Part IV
Concluding Summary of the Book

Chapter 13
General Summary on Risk Management

Abstract Risk management is designed toward getting the best value on construction project by addressing and assessing possible risks either at the onset of construction or at any phases of the project execution. In a bid to understanding the general ideology behind risk management practice in both developed and developing construction industries across selected nations, this chapter inferred that even though there are similarities in terms of concept delivery, there are subsequent differences that influence how construction professionals in both regions execute risk management practice. This constituents differences in project method, delivery, and output in relation to financial resources, technology and innovation, knowledge and expertise, and regulatory framework.

Keywords Construction risk · Project management · Project risk · Sustainable construction · Sustainable risk · Value management

Introduction

Risk management is the process of dealing with potential risks by finding, examining, and minimizing them to lower the chance or impact of harmful events and to make the most of emerging opportunities. In order to reduce the negative effects of risks in construction projects, project managers need to follow a risk management process (Construction Risk Management, 2023). This process requires careful planning to create a plan that helps project managers to spot, track, and deal with risks as they occur (Risk Management, 2023). The steps of this process include identifying possible risks, finding out who are affected by the risks, assessing the risks and making a plan to manage them. The kinds of risks that can majorly happen in construction projects are risks related to safety, cost, and materials.

A. E. Oke et al., *Risk Management Practices in Construction*,
https://doi.org/10.1007/978-3-031-35557-8_13

Definitions and Concepts of Risk Management Practices

Edward and Bowen (1998) described risk management as a helpful method to cope with the unpredictability of construction projects.

Kartam and Kartam (2001) also described risk as the possibility of some uncertain, unknown, and undesirable results changing the expected returns on a certain investment.

Dai et al. (2009) defined risk in the construction industry as an occurrence that can adversely affect the remaining goals of a project in terms of cost, time schedule, and quality.

Tohidi (2011) further explained risk management as the mental activity of identifying and evaluating risk before taking actions to lower it to a tolerable level.

Nawaz et al. (2019) expatiated on risk as an occurrence that influences the organization's objectives and that subsequently affects the organization's performance in terms of productivity, quality, and budget.

Concept of Risk Management Practices

In the construction industry, managing risks is essential for identifying, evaluating, and reducing potential risks and uncertainties that may affect a construction project. By using effective risk management techniques, construction professionals can lessen the negative impact of risks and increase the chances of project success. Academic research on risk management practices in construction offers valuable insights into different aspects of this field (Shad et al., 2019).

In construction, risk management practices involve a structured approach to identifying, evaluating, and addressing risks that may occur during various project stages such as planning, design, construction, and operation. Below are some important aspects and ideas related to risk management practices in construction (Schulte & Knuts, 2022).

Risk Identification

This refers to the process of recognizing potential risks and dangers associated with a construction project. It involves identifying both external risks, such as changes in the economy, regulations, and natural disasters, and internal risks, such as mistakes in design, delays in construction, and safety hazards.

Risk Assessment

After risks have been identified, they must be evaluated to determine the probability of them happening and their potential impact on the project. Risk assessment techniques, such as qualitative and quantitative analysis, can be used to prioritize risks and allocate the necessary resources to mitigate them.

Risk Mitigation

Strategies for risk mitigation are put in place to decrease the probability or impact of identified risks. This can include taking preventive actions, establishing safety procedures, utilizing advanced construction methods, or making changes to project plans and designs.

Risk Allocation

Different stakeholders participate in construction projects and use various methods, such as contractual clauses or insurance policies, to assign risks to the parties who can handle them better. This helps to spread risks among different parties and lower the total risk level of the project.

Risk Monitoring and Control

It is important to keep track of the risks that have been identified and control their effects throughout the project. Frequent risk assessments, performance measurements, and feedback systems help ensure that risk management plans are working well and necessary changes are made as needed.

Risk Management as a Sustainable Tool

Risk management is the practice of identifying, assessing, and mitigating risks to prevent potential losses or negative impacts on an organization or project. While traditionally focused on financial, operational, or project-based risks, sustainable risk management expands this scope to encompass environmental, social, and governance (ESG) considerations.

Risk management can be a tool for sustainable development when it considers both the threats and the benefits of sustainability issues. Organizations can use risk management as a tool for sustainability by proactively identifying and addressing risks related to environmental, social, and governance factors. This integrated approach helps ensure the long-term success and viability of businesses while promoting sustainable development. Some important aspects of this approach are discussed below as stated in Schulte and Knuts (2022) and Shad et al. (2019).

Environmental Risk

Sustainable risk management also deals with environmental risks that affect the natural environment and its resources, such as air and water pollution, climate change consequences, and loss of wildlife and habitats. This may require taking actions to lower emissions, save resources, and follow eco-friendly practices.

Social Risks

Social risks are also part of sustainable risk management, and they include those that affect human rights, working conditions, community engagement, and product quality. This could require adopting ethical business practices, ensuring respect for employees and stakeholders, and communicating with local communities.

Integration with Sustainable Strategies

Sustainable risk management is an integral part of sustainability strategies and goals. It aims to align risk management practices with broader sustainability objectives such as minimizing environmental impact, promoting social equity, and encouraging responsible governance.

Governance Risks

Governance risks are also included in sustainable risk management, and they involve corruption, dishonest practices, and low transparency. It stresses the importance of having strong corporate governance, ethical choices, and responsibility.

Long-Term Perspective

Traditional risk management often concentrates on short-term financial aspects. Sustainable risk management adopts a long-term view, taking into account the possible effects of risks and choices on future generations and the overall welfare of society.

Stakeholders' Engagement

Effective sustainable risk management requires engagement with a variety of stakeholders such as employees, communities, customers, investors, and regulators. By incorporating diverse perspectives, potential risks can be identified and addressed more comprehensively.

Lessons from Developing and Developed Countries' Risk Management Practice

The sections of the book previously discussed risk management in construction between selected developing and developed countries. Developing countries include Malaysia, Nigeria, Qatar, Saudi Arabia, South Africa, Sri Lanka, and Tanzania, while selected developed countries are Australia, Canada, Sweden, and (USA).

When comparing risk management practices in developing and developed construction industries, there are both similarities and differences. However, different countries were researched with corresponding differences inferred in terms of governing bodies, method and stage of implementation, and level of awareness. However, they all exhibit the same risk management practice with little or no differences but much similarities. After carefully researching the countries in both (developing and developed) categories, the following similarities and differences were obtained.

Similarities Between Developing and Developed Risk Management Practice

Risk Identification

The construction industry, both in developing and developed countries, recognizes the importance of identifying potential risks in construction projects. This involves conducting thorough assessments to identify risks related to safety, environmental factors, financial aspects, and project management.

Risk Assessment

In evaluating risks, construction industries in both developed and developing settings use methods to measure how likely and how serious the risks they face are. This helps them rank the risks and choose suitable ways to deal with them.

Risk Mitigation

To lessen known risks, both developing and developed construction industries strive to implement risk mitigation strategies. These strategies often include the use of safety measures, best practices, regular inspections, and appropriate technology and equipment.

Legal Compliance

The construction industries in both countries follow legal rules, building regulations, and industry norms. This makes sure that projects achieve the needed safety and quality levels, lowering the chance of accidents, setbacks, and legal problems.

Differences Between Developing and Developed Risk Management Practice

Financial Resources

Construction industries in more developed countries often have financial capability to spend on managing risks than those in less developed countries. This lets them use advanced technology, extensive insurance plans, and expert risk management teams. Construction industries in developing countries may have limited resources, which can affect the quality and complexity of their risk management practices.

Technology and Innovation

Construction companies (industries) in developed countries often have greater access to advanced technology and innovative solutions for risk management. This includes tools like Building Information Modeling (BIM), drones for site inspections, and real-time monitoring systems, which can enhance risk detection, evaluation, and response. In contrast, construction companies in less developed countries may have limited access to such technologies, which can affect their ability to manage risks effectively.

Knowledge and Expertize

Construction industries in developed countries often have the advantage of a well-established knowledge base, skilled professionals, and dedicated institutions for risk management. This allows them to leverage extensive industry experience, best practices, and research findings. In contrast, construction industries in developing countries may face challenges in transferring knowledge, developing skills, and accessing relevant expertise.

Regulatory Framework

The regulatory environment and enforcement mechanisms can vary between developed and developing countries. Developed countries often have more extensive and strictly enforced regulations related to construction safety, environmental impact, and labor practices. In contrast, developing countries may be in the process of improving their regulatory frameworks or may face challenges in consistently enforcing regulations.

Effective Risk Management for Construction Project

Construction projects are naturally complicated and involve many risks and uncertainties that can affect project objectives. Managing these risks effectively does not mean getting rid of them completely, which may seem like the cheapest option but it is not the case every time since every project is different. From an economic perspective, the approach of complete elimination of risks is futile because profitable activities naturally involve risk, and activities without any risk are not economically feasible or beneficial (Wróblewski, 2007). However, Kwansoo and Chavas (2003) argue that effective risk management is essential for a good management strategy. A good risk management plan helps an organization to take more risks while increasing its overall safety. The different ways of dealing with the possibilities of losses include avoiding, reducing, transferring by contract, and transferring by insurance and keeping. Usually, the best way to manage risks is to use more than one methods together. It is important to have skilled coordination in applying the selected risk treatment methods to achieve significant improvement and measures results accurately. Sharma and Swain (2011) also emphasized that project management has changed over time, and managing risks in project delivery has become very important. Project teams now try to meet customer and end-user needs, and avoid common problems in construction projects such as delays, cost overruns, low quality, and unsafe work environments. The key to achieving these goals is to manage risks effectively throughout the project lifecycle. Using a holistic risk management approach helps organizations to identify all the business risks they face, increasing the chance of successful risk reduction and, ultimately, complete risk elimination. While stressing the need to manage risks using a holistic approach, Labbi (2005) stated that risk factors are not independent but interrelated. If we ignore their connections, we may end up managing risks in a wasteful or ineffective way. For example, we may hedge some risks that are actually balanced by others or we may miss some important risks that could harm the whole organization. We should look at the big picture of how risks affect the organization and avoid actions that could create new risks or shift them to other parts of the organization.

Lam (2003) further claims that managing risks using a holistic approach has three main benefits. First, it makes the organization more effective by aligning the top and bottom levels, promoting cooperation among different functions, and dealing with not only single risks but also their connections. Second, it enhances risk reporting by assigning accountability and setting uniform reporting standards. This allows the consolidation of risks to give a complete picture of risks throughout the whole enterprise. Third, organizations that use this approach have seen significant improvements in business performance. Holistic risk management helps key management decisions such as how to allocate capital, develop and price products, and merge or acquire other businesses. As a result, it leads to lower losses, less earnings variability, higher earnings, and increased shareholder value.

In general, effective risk management is essential to identifying, evaluating, and reducing risks, thereby ensuring successful project completion. The key elements of effective risk management in construction projects are summarized below.

Risk Identification

In construction projects, the process of identifying risks mainly involves understanding the kinds of risks that could potentially affect the project. This involves identifying the specific factors that describe these risks and estimating their probability of happening within the project. The need for risk identification comes from the decision-making situations faced by an investor at any given time (Bizon-Górecka, 2006). Here, there is need to consider various factors such as design difficulty, site situations, regulatory compliance, contractual requirements, stakeholder expectations, and external influences. Techniques like brainstorming, checklists, historical data analysis, and expert opinions help in comprehensive risk identification.

Risk Analysis

There is need to analyze identified risks to understand their likelihood of happening and potential effect on project goals. Risk analysis can use quantitative or qualitative methods. Quantitative analysis involves using a number to measure the risk level. Various techniques can be used in quantitative risk assessment, such as expected value analysis, decision trees, probability analysis, fault trees, event trees, sensitivity analysis, risk simulation, and statistical inference. However, using only quantitative measures to determine risk levels has some problems. Firstly, accurate data is very important for measuring risks correctly, requiring careful attention to data accuracy. Secondly, the risks that can be measured quantitatively depend on the availability of relevant data. Thirdly, many risk factors such as political, personnel, and regulatory risks, as well as intangible consequences like reputation damage and negative media coverage, are hard to measure with quantitative techniques. However, these factors can expose organizations to significant risks (Schwartz & Smith, 1997). This analysis helps rank risks and focus resources on managing crucial ones.

Risk Evaluation

This process evaluates the importance of identified risks to make informed decisions on risk response strategies. According to ISO 31000:2009, risk evaluation is a process that compares the results of risk analysis with predefined risk criteria established during the context setting stage. The purpose is to determine whether the identified risks needs treatment and to prioritize the implementation of treatment measures. Alternatively, further analysis may be needed to make informed decisions. The risk criteria are based on the desired risk profile of the organization which reflects its risk appetite. This refers to the level of risk that the organization is willing to take in its pursuit of value, often representing a balance between growth, risk, and return. Risks that fall within the established limit are considered acceptable and require only monitoring without specific treatment. On the other hand, risks considered unacceptable will require treatment measures. Here there is need to consider project objectives, constraints, stakeholder priorities, and risk tolerance levels, and to evaluate risks based on their potential impact on cost, schedule, quality, safety, and overall project success. This evaluation assists in determining whether to accept, mitigate, transfer, or avoid risks.

Risk Treatment

This process involves developing and implementing strategies to lower the probability or effect of identified risks. ISO 31000 emphasizes that risk treatment involves the selection and implementation of options to modify risks, with these options not being mutually exclusive. Once applied, treatments create or change controls. The process involves identifying, evaluating, preparing, and implementing a range of options for addressing risks (Knight, 2010). Treatment options may include (1) accepting the risk, (2) avoiding the risk, (3) lowering the risk, and (4) sharing or transferring the risk. Moeller (2007) suggests that organizations should carefully consider the costs and benefits of each potential risk response. It is important to align these strategies with the organization's overall risk appetite when developing a risk strategy for each specific risk. Construction firms can implement preventive measures like improved design, strong construction techniques, strict quality control, safety protocols, contractual risk allocation, and contingency planning. Regular monitoring, inspections, and audits help ensure effective implementation of risk reduction measures.

Risk Monitoring and Review

The probability and effects of an outcome can be affected by various factors, and these factors have the potential to change over time. Likewise, the factors that impact the appropriateness or cost of different treatment options can also undergo changes (Hintsay, 2016). Therefore, there is need to continuously monitor and review risks throughout the project lifecycle. Regularly evaluating any changes in project execution, external influences, stakeholder preferences, and market trends will help in managing risk associated with such project. Risk registers, performance indicators, progress reports, and feedback mechanisms to track risks and update risk management strategies are also to be taken into account.

Risk Communication

This process involves ensuring that everyone involved in the project knows about the risks and how to deal with them. Schieg (2006) highlighted that, in implementing effective risk management in an organization, it is important to have good information that allows complete and informed decision-making. This means that the organization needs to communicate well so that the right stakeholders get the right information so as to prevent irrational responses which are vital to avoid possible losses or the failure to achieve expected benefits (Sousa et al., 2012). This process encourages teamwork and cooperation among project staffs, contractors, subcontractors, suppliers, and workers to make clear decision on what they need to do to manage risks. This can be achieved by sharing risk-related information regularly through meetings, reports, training sessions, and documentation.

Benefits of Effective Risk Management to Project Team

Effective risk management in construction projects has many benefits, both in the short and long terms. Implementation of effective risk management helps to deal with possible risks that could affect the main objectives of the project, such as budget, schedule, and quality. Some of the benefits to project team are highlighted below.

Improved Project Planning

Effective risk management allows project managers to evaluate possible risks and create detailed plans that accounts for different scenarios. It makes sure that potential risks are spotted and handled early, lowering the chance of costly problems during project implementation, such as delays, defects, accidents, and disputes (Hintsay, 2016).

Better Decision-Making

Hintsay (2016) identified that, effective risk management will make an organization's operation more efficient and minimize reduction to normal operations, handle changes better, and choose the best options when making decisions. By carefully examining risks, project stakeholders learn more about the possible effects and outcomes of different decisions. This enables them to make wise choices and pick the best options based on their risk tolerance and project goals.

Budget and Schedule Control

Effective risk management helps the project team to identify possible cost or time overruns before they occur. This lets them act quickly by applying suitable measures to reduce these risks. This improves the reliability and accuracy of project projections, keeping budget changes low, and making sure the project meets its deadlines.

Improved Stakeholders Communication

Construction projects involve stakeholders with different views and attitudes toward risk. Risk communication and consultation helps to identify, record, and understand stakeholders' perceptions, as it is also useful for supporting future applications of the risk management processes (Cooper et al., 2005). Effective risk management promotes open and proactive communication among project stakeholders. It helps to identify and understand risks by all parties involved, ensuring collaboration, trust, and agreement toward shared project objectives.

Improved Resource Allocation

Hopkin (2012) argue that useful risk management information helps to make better decisions about strategic and operational options and supports effective use of limited resources by weighing risks and benefits of different strategies. However, by managing risks effectively, project teams can use their resources better by focusing on the most important risks. This helps them avoid wasting resources and improve their project results.

Stakeholder Confidence and Organization's Reputation

When construction projects take steps to identify and deal with risks before they become problems, they show that they care about delivering what they promised and protecting the people who are involved or affected by the project. This makes the clients, investors, and the public trust and respect them more for their professionalism and responsibility. This also protects the project from getting a bad reputation and helps the organization build a good image for itself.

Achieving Sustainable Projects

Effective risk management is to ensure that construction projects can succeed in the long run by avoiding or reducing potential problems. It makes sure that risks are dealt with properly and in a timely manner throughout the project, making it more likely to achieve project objectives and deliver good results.

Summarily, effective risk management in construction projects enables project teams to plan better, make smarter decisions, keep costs and schedules under control, communicate better, use resources more efficiently, build trust with stakeholders, and achieve project objectives in the long run. By taking action on risks before they become problems, construction organizations can deal with uncertainties better and improve their chances of project success.

Benefits of Effective Risk Management to Clients

Sousa et al. (2016) postulated that it is essential that the project owner and their representatives start the risk management process. This way, they can make sure that all stakeholders are taken into account and that the project meets the end user's needs. Then, designers, contractors, and other parties should follow the owner's risk management instructions to achieve the end user's objectives. Effective risk management is crucial for achieving successful results for clients in construction projects. By identifying, evaluating, and reducing risks, it offers many benefits that improve project delivery, prevent potential losses, and safeguard the clients' interests. The following are main benefits of effective risk management to clients in construction projects.

Cost Control

Meulbroek (2002) stated that, effective risk management helps to avoid the usual problem of having separate risk management teams that are inefficient and in conflict with each other because they do not coordinate well. This way, risk can be managed more effectively and at a lower cost. Effective risk management will help clients to keep project cost under control and within budget. This can be achieved by identifying the risks early and then take action to reduce or avoid them. This lowers the chances of unforeseen costs, budget overruns, and financial losses. By addressing risks proactively, clients can better foresee and handle financial risks, making sure that the project stays within budget limits.

Schedule Adherence

Finishing the project on time is very important to clients. Effective risk management helps to identify and reduce risks that may slow down the project, by planning and dealing with possible schedule problems, clients can lower the chance of project delays and extra costs. This makes sure that the project is completed on time, meeting client expectations, and avoiding any negative effects on business operations or future plans (Bizon-Górecka, 2006).

Quality Assurance

Clients in construction projects expect the quality of the project to meet or exceed their standards and specifications. To ensure this, project stakeholders need to manage risks well throughout the project lifecycle. This involves identifying and managing possible risks that could compromise the quality of the project, such as design errors, material defects, or construction faults. By applying strong quality control methods, such as inspections, tests, audits, or reviews, they can monitor and verify the quality of the project at every stage. By dealing with risks that may affect the project's outcome, such as changes in scope, budget, or schedule, project team can prevent or minimize any negative impacts on the quality of the final product. By doing so, clients can have greater confidence in the final product and its compliance with their requirements and expectations. This enhances client satisfaction with the project delivery and performance. It also protects their reputation as a reliable and trustworthy partner in the construction industry. Moreover, it boosts the value and longevity of the constructed facility by ensuring its functionality, safety, and reliability.

Stakeholder Communication

Risk management requires clear and honest communication among all project stakeholders. Managing risks well makes sure that clients are aware of potential risks, how to deal with them, and how the project is going. Frequent communication and engagement build trust, encourage joint decision-making, and let clients take part in risk-related conversations. Clients can offer useful feedback, help with risk reduction efforts, and match project objectives with their own organizational goals.

Improved Overall Project Success

Managing risks effectively greatly enhances the overall success rate of construction projects. By taking action on risks before they become problems, clients can lower the chance of project failures, conflicts, and unforeseen outcomes. This results in higher project success rates, happier clients, and positive project outcomes that match the client's vision and objectives. Effective risk management is a key factor that boosts the overall success rate of construction projects.

Safety and Risk Mitigation

Effective risk management helps organization to have a full understanding of the risk involved in its activities, what causes it, and how to manage it well with the right strategies to reduce it (Meulbroek, 2002). Construction sites pose a lot of risks to workers, the public, and the environment. Effective risk management aims to find and reduce potential safety and environmental problems. By putting safety first, clients can create a safe working environment, prevent accidents, and avoid possible legal and financial problems. This not only keeps everyone safe but also preserves the client's reputation and prevents expensive legal conflicts.

Risk Management and Sustainable Construction

Sustainable construction has been a major focus for construction players across the globe in recent times. Kibert (2016) described sustainable construction as a way of infrastructure development which involves the planning, building, and running of green building projects that also aims to minimize negative environmental impacts, conserve resources, promote social responsibility, and ensure long-term economic viability. Robichaud and Anantatmula (2011) in their research on green constructions identified that the increase in traditional building material costs and energy prices has prompted individuals and organizations to consider sustainable construction as a viable solution. Government incentives and regulations have further amplified this trend by providing additional motivation and support. Together, these factors have fostered a shift toward sustainable construction practices that prioritize resource efficiency, environmental responsibility, and long-term economic viability. The researchers further acknowledged that, despite the development of green construction, there are still some risks that needs to be addressed; one big worry is the possible failure to deliver projects within acceptable cost limits. Green construction often means adding sustainable features and technologies that may cost more at first than conventional options. This cost difference can stop some developers or clients from going for sustainable construction projects, especially if they have tight budget restrictions. Finding a balance between the financial aspects of a project and adding sustainable elements need careful planning, cost analysis, value engineering, and effective risk management to make sure that the project is still economically viable. Vitunskaite et al. (2019) also agrees that new methods and ideas in sustainable construction, especially in the area of smart cities, can bring more risks because they are relatively new and not well understood by project management teams. When using new technologies or unusual practices, there may be doubts and risks that come from the lack of previous experience or proven best practices. Vitunskaite et al. (2019) acknowledges that this inherent risk is more common in the new concept of sustainable smart cities, where the combination of advanced technologies and complex systems is still changing.

In sustainable construction, it is important to identify and evaluate possible risks in order to manage them well and make sure the project succeeds (El-Sayegh et al., 2018). Risk identification means finding all possible risks and uncertainties that could affect the project's goals, schedule, cost, quality, and sustainability goals. This process needs to involve stakeholders with different skills and views, such as architects, engineers, contractors, and sustainability consultants. After identifying risks, they need to be evaluated based on how likely they are to occur and how much they could affect the project. This evaluation helps rank risks based on how important they are and lets project teams use the right resources for risk response planning. Risks in sustainable construction projects can be different, from technical problems related to green technologies to regulatory compliance, market changes, and resource availability. By evaluating risks well, project teams can get a good understanding of the potential problems and opportunities related to sustainable construction projects.

After risk evaluation, the next step is risk response planning and control. This means making strategies and actions to reduce or get rid of risks, transfer risks through insurance or contracts, accept some risks, or make backup plans. Risk response planning tries to lower the chance and effect of found risks while improving project results. It may mean changing some parts of the project, using different materials or technologies, having more suppliers, or making backup plans for getting resources. Risk control is a continuous process throughout the project. It means checking and reviewing how well risk response strategies work and making changes as needed. Regular communication and cooperation among project stakeholders are important for good risk control, as it allows for the quick finding of new risks and the changing of risk response plans as needed. Moreover, identifying and evaluating risks in sustainable construction projects in advance, project teams can avoid potential problems, make sure project objectives are achieved, and improve the overall success and sustainability of the construction project.

Conclusion

This chapter discussed the general summary of the previous chapters, starting from the general introduction of the topic (risk management) and extending toward definitions relating to the topic. Furthermore, this chapter discussed the similarities and differences experienced in the selected regions of the research (developed and developing nations). By assessing the uniqueness of both regions, the research was able to make adequate comparison between the categorized regions selected and gave an exclusive inferences. Finally, delving into the benefits of risk management to both the client and the project team, coupled with the inclusiveness of risk management in achieving sustainable construction, this chapter gave an overview of the benefits of assessing risks in terms of managing construction, enhancing project delivery, and reducing cost, which correlates with the concept of achieving a sustainable construction industry.

References

Bizon-Górecka, J. (2006). Innovative applications in the construction industry. *Scientific Papers of the Institute of Building Engineering, 87*(2), 175–182.

Construction Risk Management. (2023). *An introduction to project management.* Retrieved from https://www.projectmanager.com/blog/construction-risk-management

Cooper, D. F., Grey, S., Raymond, G., & Walker, P. (2005). *Project risk management guidelines. Managing risk in large projects and complex procurements*. Wiley.

Dai, J., Goodrum, P., Maloney, W., & Srinivasan, C. (2009). Latent structures of the factors affecting construction labour productivity. *Journal of Construction Engineering and Management, 135*(5), 397–406. https://doi.org/10.1061/(ASCE)0733-9364(2009)135:5(397)

Edwards, P. J., & Bowen, P. A. (1998). Risk and risk management literature in construction: A review and future directions for research. *Engineering, Construction and Architectural Management, 5*(4), 339–349.

El-Sayegh, S. M., Manjikian, S., Ibrahim, A., Abouelyousr, A., & Jabbour, R. (2018). Risk identification and assessment in sustainable construction projects in the UAE. *International Journal of Construction Management, 1*(1), 1–11. https://doi.org/10.1080/15623599.2018.1536963

Hintsay, H. (2016). *Integrated risk management practice in the construction industry general contractors' perspective.* Masters dissertation, Department of Management, Addis Ababa University College of Business and Economics, Ethiopia.

Hopkin, P. (2012). *Fundamentals of risk management: Understanding, evaluating and implementing effective risk management.* Kogan Page Publishers, London.

International Standard Organization ISO. (2009). *ISO 31000:2009, risk management—Principles and guidelines*. International Standards Organization.

Kartam, N., & Kartam, S. (2001). Risk and its management in the Kuwaiti construction industry: A contractors' perspective. *International Journal of Project Management, 19*(6), 325–335.

Kibert, C. (2016). *Sustainable construction: Green building design and delivery* (4th ed.). Wiley.

Knight, K. W. (2010). Risk management: A journey not a destination. In *Presentation to the Rusrisk/Marsh—ISO 31000 risk management standard: Principle and implementation trends seminar*, 15th December, Moscow.

Kwansoo, K., & Chavas, J. P. (2003). Technological change and risk management: An application to the economics of corn production. *Agricultural Economics, 29*(2), 125–142.

Labbi, A. (2005). *Handbook of integrated risk management for e-business*. Ross Publishing.

Lam, J. (2003). *Enterprise risk management: From incentives to controls*. Wiley Finance Series.

Meulbroek, L. K. (2002). Integrated risk management for the firm: A senior manager's guide. *Journal of Applied Corporate Finance, 14*(1), 56–70.

Moeller, R. R. (2007). *COSO enterprise risk management: Understanding the new integrated ERM framework*. Wiley.

Nawaz, A., Waqar, A., Shah, S. A., Sajid, M., & Khalid, M. I. (2019). An innovative framework for risk management in construction projects in developing countries: Evidence from Pakistan. *Journals of Risks, 7*(1), 24–33. https://doi.org/10.3390/risks7010024

Risk Management. (2023). *Risk management in construction: Process of managing risk*. Retrieved from https://www.stakeholdermap.com/risk/risk-management-construction.html

Robichaud, L., & Anantatmula, V. (2011). Greening project management practices for sustainable construction. *Journal of Management Engineering, 27*(1), 48–57.

Schieg, M. (2006). Risk management in construction project management. *Journal of Business Economics and Management, 7*(2), 77–83.

Schulte, J., & Knuts, S. (2022). Sustainability impact and effects analysis - A risk management tool for sustainable product development. *Sustainable Production and Consumption, 30*(3), 737–751.

Schwartz, R. J., & Smith, C. W. (Eds.). (1997). *Derivatives handbook: Risk management and control (Vol. 6)*. John Wiley & Sons, New Jersey, USA.

Shad, M. K., Lai, F., Fatt, C. L., Kleme, J. J., & Bokhari, A. (2019). Integrating sustainability reporting into enterprise risk management and its relationship with business performance: A conceptual framework. *Journal of Cleaner Production, 208*(2), 415–425.

Sharma, S. K., & Swain, N. (2011). Risk management in construction projects. *Asia Pacific Business Review, 7*(3), 107–120. https://doi.org/10.1177/097324701100700310

Sousa, V., De Almeida, N. M., & Dias, L. A. (2012). Risk management framework for the construction industry according to the ISO 31000:2009 standard. *Journal of Risk Analysis and Crisis Response, 2*(4), 261–274.

Sousa, M. R., Gama, J., & Brandão, E. (2016). A new dynamic modeling framework for credit risk assessment. *Expert Systems with Applications, 45*(1), 341–351.

Tohidi, H. (2011). The role of risk management in information technology systems of organizations. *Procedia Computer Science Journal, 3*(1), 881–887.

Wróblewski, P. (2007). *Information technology project management for practitioners*. Helion Publishing.

Index

A. E. Oke et al., *Risk Management Practices in Construction*,
https://doi.org/10.1007/978-3-031-35557-8

R

S

T

U

V

W